BURLEIGH DODDS SCIENCE: INSTANT INSIGHTS

NUMBER 85

Improving the health and welfare of heifers and calves

bd burleigh dodds
SCIENCE PUBLISHING

Published by Burleigh Dodds Science Publishing Limited
82 High Street, Sawston, Cambridge CB22 3HJ, UK
www.bdspublishing.com

Burleigh Dodds Science Publishing, 1518 Walnut Street, Suite 900, Philadelphia, PA 19102-3406, USA

First published 2023 by Burleigh Dodds Science Publishing Limited
© Burleigh Dodds Science Publishing, 2024. All rights reserved.

British Library Cataloguing in Publication Data
A catalogue record for this book is available from the British Library

ISBN 978-1-80146-643-1 (Print)
ISBN 978-1-80146-644-8 (ePub)

DOI: 10.19103/9781801466448

Typeset by Deanta Global Publishing Services, Dublin, Ireland

Contents

Series list

Title	Series number
Sweetpotato	01
Fusarium in cereals	02
Vertical farming in horticulture	03
Nutraceuticals in fruit and vegetables	04
Climate change, insect pests and invasive species	05
Metabolic disorders in dairy cattle	06
Mastitis in dairy cattle	07
Heat stress in dairy cattle	08
African swine fever	09
Pesticide residues in agriculture	10
Fruit losses and waste	11
Improving crop nutrient use efficiency	12
Antibiotics in poultry production	13
Bone health in poultry	14
Feather-pecking in poultry	15
Environmental impact of livestock production	16
Sensor technologies in livestock monitoring	17
Improving piglet welfare	18
Crop biofortification	19
Crop rotations	20
Cover crops	21
Plant growth-promoting rhizobacteria	22
Arbuscular mycorrhizal fungi	23
Nematode pests in agriculture	24
Drought-resistant crops	25
Advances in detecting and forecasting crop pests and diseases	26
Mycotoxin detection and control	27
Mite pests in agriculture	28
Supporting cereal production in sub-Saharan Africa	29
Lameness in dairy cattle	30
Infertility and other reproductive disorders in dairy cattle	31
Alternatives to antibiotics in pig production	32
Integrated crop–livestock systems	33
Genetic modification of crops	34

Chapter 1

Optimising reproductive management to maximise dairy herd health and production

Norman B. Williamson, Massey University, New Zealand

1 Introduction

The domestication of cattle has been linked with the beginnings of civilization over 10 000 years ago. Keeping cattle for both milk and meat production occurs worldwide. Herds vary in size from one or two cows to many thousands under single management. Two main types of management systems prevail in commercial dairy farms. In one system cattle are housed in buildings or feedlots and are fed rations that are brought to them. Such systems are

http://dx.doi.org/10.19103/AS.2020.0086.09

prevalent in North America and South America, Europe, the Middle East and Asia. The other system has herds living on pasture which they graze for most of their sustenance. Grazing systems occur commonly in Australia, New Zealand, Ireland, parts of Argentina and Chile. These two systems require different management strategies for reproduction that lead to different approaches to breeding and different limits to reproductive health and performance. However, there are similarities in the fundamental principles that apply under both management systems.

Efficient dairy farms produce high volumes of product for the resources employed. Products include milk, male calves for sale, female calves or heifers not required for production, cull cows and sometimes production cows. Farms may also sometimes sell excess feed and, in developing countries, manure can also be a product that is traded. Profitable farms have a large positive difference between income, made up of the amounts of product multiplied by the prices received for them, and costs, which comprise the types of resources used multiplied by their costs.

Reproduction has a major influence on dairy farm efficiency, economics and profitability. Cows calve as a result of mating, insemination or transfer of an embryo. Most modern dairy farms in advanced economies now employ artificial insemination to improve herd genetics and selection of cows for production gain. Calving produces calves that can either be sold or enter the milking herd as replacement cows that produce milk, given that calving initiates lactation (milk production) by the cow.

The interval between consecutive calving events influences milk production efficiency. Shorter calving intervals mean that cows spend relatively more of their milking time in the earlier stages of lactation where production is higher and more efficient. They also produce more calves in a period. Efficient and adequate reproduction provides more offspring that enable discretionary culling of low-producing animals. Increased numbers of replacement heifers enable the retention in the herd of replacements with higher genetic selection differentials which leads to more rapid genetic gain. Efficient reproduction by heifers enables them to join the herd sooner and this limits the overhead costs of rearing. In pasture-based dairy production systems, seasonal calving is often managed so that calving occurs in a restricted period of the year just before peak pasture growth occurs because that allows the requirements of cows for feed to be matched to the available pasture growth.

Inadequate reproduction increases costs and may require animals to be culled. Such enforced culling requires cows to be removed from the herd before they reach their maximum milk production potential, normally at the sixth or seventh lactation, which limits the efficiency of production. Enforced culling also demands more replacement animals, thus reducing the ability of farmers to select only the most genetically superior animals.

Reproductive limits apply at an individual cow level due to variations in reproductive health and physiological performance of individuals. There are also herd reproductive limits associated with approaches to herd breeding, feeding, health and general management.

Reproductive diseases and limitations can occur at all stages of the reproductive cycle, from the onset of puberty until an animal is culled or dies. A failure of or delay in the display of oestrus can occur at expected puberty or after a previous calving. This limits the opportunities for heifers and cows to be mated and delays breeding until oestrus occurs under natural conditions. If oestrus does occur, cows and heifers may not be bred and so the opportunity for reproductive success is delayed. If breeding does occur, conception may or may not occur. If not, a successful reproductive outcome is again delayed. After conception, a pregnancy may terminate in early embryonic death and may not establish. If a pregnancy is established, it can then be aborted. If gestation is completed pregnancy may end in a dystocia or stillbirth and may be followed by a retained foetal membrane.

In the reproductive cycle of cows, these critical events enable reproductive success and are the events about which data must be collected in herd records to allow performance to be monitored and analysed. Data collection and analysis are now normally kept as a part of a computer-based recording and analysis system. Such systems greatly aid the collection and analysis of records that can be used to monitor the performance of individual cows and whole herds of many thousands of cows. Reproductive management using records to detect and identify reproductive limits to production is applied to minimise reproductive disease and reduce the impact of disease and inadequate management on reproductive performance (Williamson, 1987). This can inform interventions that optimise reproductive efficiency, health, performance and productivity (Fetrow et al., 1990).

2 Grouping animals to measure individual animal reproduction limits

Reproductive efficiency in a herd can be managed by defining the parameters of expected reproductive performance and selecting cows that fail to meet these parameters. This leads to cows being examined in defined groups during planned reproductive health visits by veterinarians. These groups include:

- Anoestrous cows;
- Nymphomaniacs;
- Post-partum checks;
- Abnormal discharges;

- Pregnancy checks; and
- Pregnancy rechecks.

These are discussed in the following sections.

2.1 Anoestrous cows

Failure to display or detect oestrus is a major limit to reproductive efficiency in most dairy regions. Cows that have not shown oestrus within 49 days of calving or those that have not shown a return to a previous unbred oestrus within 35 days (this time can be set at the lower limit of pregnancy diagnosis accuracy) should be selected for examination and treatment in year-round calving herds. These are called NVO for no visible oestrus cows or ONO for oestrus not observed cows. When selected, these cows can be examined and treated appropriately to ensure a rapid return to normal oestrus cycling. If they have a corpus luteum (CL) present (indicating prior-ovulation) they may be injected with prostaglandin to induce a rapid onset of oestrus. Otherwise they may be started on an oestrus induction programme such as one of the progesterone, prostaglandin, gonadotrophin programmes to induce a return to normal oestrus cycling. The presence of a CL also indicates that a previous oestrus is likely to have occurred and to not have been observed or recorded. This indicates that aids to oestrus detection such as dye-containing pressure-sensitive patch detectors, tail painting, activity monitoring or pressure-activated electronic oestrus detection devices should be considered to improve the sensitivity of oestrus detection.

Cows and heifers that are not pregnant at the normal period for pregnancy diagnosis de facto join this group and can be treated in similar ways. Cows in this situation are sometimes called 'phantom' cows. An analogous approach can be taken with heifers that have not demonstrated oestrus by the desired time of first breeding. All heifers not detected in oestrus by 13 or 14 months of age can be examined and treated in a similar way to the NVO cows.

2.2 Nymphomaniac cows

If oestrus is observed frequently, that is, three recorded occurrences of oestrus within 36 days, in the absence of treatment with an oestrus cycle shortening drug such as prostaglandin or an oestrogen, cows displaying these signs should be examined so that they can be checked for cystic ovarian disease and appropriately treated.

2.3 Post-partum check cows

Post-partum checks are required on cows that have a previous abnormality associated with calving. Abnormalities include induced calving, dystocia, prolapsed uterus, retained foetal membranes and post-partum metritis. These cows are routinely examined from about 20 days after calving even if no external abnormal signs are evident. Some veterinarians, particularly in the northern hemisphere, examine all cows routinely during post-partum or pre-breeding examinations in order to detect clinical or sub-clinical endometritis (Dubuc et al., 2010). This selection for examination became more popular after the introduction of the 'Metricheck' instrument that enables mucus and any pus it contains to be extracted from the cranioventral vagina near the cervix and scored as to the likelihood of endometritis. Cows deemed to have endometritis can then be treated with an antibiotic, commonly cephapirin which is not absorbed parenterally or excreted in the milk. Any cow that has clinical signs of metritis or that has an abnormal vaginal discharge should also be selected to be examined and treated if required.

2.4 Cows with abnormal discharges

If abnormal, usually purulent, discharges from the vulva are seen, cows should be selected for examination and treated if required.

2.5 Pregnancy check cows

All cows that have been inseminated or bred, and which have not been detected in oestrus for a time that exceeds the lower limit for accurate pregnancy diagnosis (often set at 35 days), should be selected as a pregnancy diagnosis (PD) category. It is a useful practice to confirm and estimate the stage of the pregnancy without knowing the anticipated stage of pregnancy. This is so any discrepancy between the determined and expected stages of pregnancy can be considered and the most appropriate stage of pregnancy allocated.

2.6 Rechecks for pregnancy

Rechecks for pregnancy are required if cows that previously were diagnosed pregnant have oestrus or insemination recorded after the positive PD. Rechecks can also be requested by a veterinarian if a pregnancy appears to be abnormal or if it is too difficult to diagnose with confidence at an initial attempt. Cows that exceed their expected calving date by 3 weeks or longer should also be re-examined for pregnancy. Where abortion has occurred, tests to determine possible causes should be conducted.

Pregnancy diagnosis is essential to determine the status of individual cows so they may be managed appropriately. Accumulated pregnancy status information is required to allow key herd reproductive indices to be calculated and to facilitate monitoring, including conception rates and calving to conception intervals.

3 Measuring reproductive performance

Despite easy recognition of reproductive inadequacy in herds, achievement of improvement requires a thorough understanding of reproductive processes and relationships in order to specify biological or management reasons limiting reproduction. Farmers and veterinarians frequently recognise that improvements in herd reproductive performance are needed when performance achieved in herds does not match targets.

Reproduction is managed using measures or indices that allow performance to be recorded and then analysed. These performance measures allow targets to be set so that herd reproductive performance can be monitored by comparing the performance achieved against these targets. They can thus help identify inadequate reproductive performance. Where this occurs, some reproductive indices also help to identify likely causes of inadequate reproductive performance.

Indices may be used to measure performance of individual animals or herds. Some indices of reproduction are directly related to production, provide easy measures of farm efficiency and a way to detect that there are problems occurring on a farm, for example, long calving intervals or high culling rates. However, such indices provide no useful information about the causes of problems. Other indices are more focused and reflect performance in specific areas of farm reproductive performance that offer information on what is limiting reproductive performance in a way that allows diagnosis. These can be characterised as diagnostic indices.

To document reproductive performance, complete and accurate herd records are required to be maintained. These records may be found in a variety of forms such as wall sheets, cow cards or books. However, they are more commonly now kept as computer-based records either on the farm, for example, in systems provided by milking equipment companies, or kept centrally by herd test or artificial breeding organisations.

Records consist of details of cows including identification, birth date, the dates and details of all calving events, dates of all observed oestruses and breeding dates and details. Breeding events should record dates, sires or semen used and, when artificial insemination is used, the inseminator's identification. If natural breeding is used and a bull or bulls are run with the herd, the date when bulls are introduced to the herd and date removed should also form a part of the records. If a herd is under veterinary supervision through an ongoing herd reproductive health programme, the dates and findings at all reproductive examinations should form a part of the record. Data recorded will include the pregnancy status and stage, the date and nature of any reproductive abnormality, the dates of all treatments and details of the products used, dry-off dates and the date and reasons for removal from the herd.

Measures of reproductive performance include intervals measured in days which can highlight a potential issue where a discrepancy from target occurs. When reproductive intervals are being recorded and evaluated, it is simple to calculate and consider mean intervals. However, this is not always appropriate as some reproductive intervals have multimodal distributions and are better represented as frequency distributions.

4 Production-related reproductive indices for pasture-based seasonally calving herds

In seasonally calving herds, the time and distribution of calving are related to the production of offspring, sale animals and milk and are thus production-related indices. Production-related reproductive indices include:

- Four- and 8-week calving rates;
- Mean calving date and planned start of calving date;
- Proportion of cows calving that were induced (where it is allowed);
- Proportion of calved cows that are culled for inadequate reproduction; and
- Proportion of cows carried over for a season that are non-pregnant.

If these indices fail to be on target, they indicate a potential problem but provide little information on why reproductive performance is not meeting targets. These measures are production related as they influence the amount and efficiency of output of milk and livestock through early compact calving. They do not, however, provide much information on causes or specific factors that lead to inadequate performance. It is known, however, that calving pattern in one season is highly predictive of the calving pattern in the subsequent season.

The season is defined by the planned start of calving date. This sets the desired date for calving to start. The spread of calving around that date is usefully displayed graphically as a frequency distribution of the cows calving around that date. This can be presented as a graph (e.g. Fig. 1).

The InCalf programme as used in New Zealand reports the percentage of cows calved by 3, 6 and 9 weeks after the nominated planned start of calving date and this helps to describe performance in the calving season. Top farmers achieve 60%, 87% and 98%, respectively, for these indices (Fig. 2).

This means of evaluating reproductive performance can be distorted by management interventions such as by the pharmacological induction of late calving cows where this is permitted. If it is permitted, then the proportion of the herd induced to calve early should be recorded and monitored to evaluate the influence of this intervention on the calving pattern. The spread of calving also can be influenced by culling of cows that do not conceive within the planned

```
DairyCHAMP : Database Applications                         Farm: MCDONALD
- - - - - - - - - - - - - - - - - - - - - - - - - - - - - - - - - - - - - - -
EVENTTALLY
```

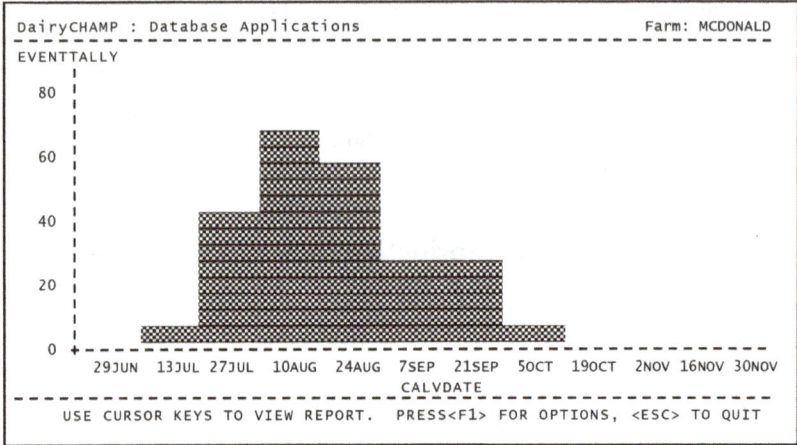

Figure 1 Histogram of number of cows calving by time.

mating period and that are likely to calve outside the planned season. This can be measured as the empty (non-pregnant) rate at the end of the mating period. Cows and heifers that do not conceive within the mating period and that are then culled influence the subsequent calving spread because they do not figure in it. This makes it important to monitor and report the proportion of cows mated with the intention of having them calve in the next calving season that are culled for reproductive failure because they did not become pregnant in time. This measure includes cows that do not conceive and those that conceive too late to calve in the following defined calving season.

5 Diagnostic reproductive indices for pasture-based seasonally calving herds

For seasonally calving herds in Australia and New Zealand, a key index of performance relating to the breeding season that is used by farm advisors

Calving pattern of whole herd Did late calvers reduce in-calf rates?			
Calved by	Week 3	Week 6	Week 9
Your herd	47%	78%	95%
Aim above	67%	88%	98%
	⬆	⬆	⬆

Figure 2 Calving distribution table from the InCalf Fertility Focus Users Guide (2020).

(including veterinarians) is the 6-week in-calf rate. That is the proportion of cows present in the herd at the planned start of mating that are bred and/or inseminated in the first 6 weeks of mating and are diagnosed to be pregnant.

Reproductive diagnostic indices pertinent to seasonally calving herds include the submission rate and the conception (pregnancy) rate or risk. In seasonal dairy herds where conception occurs in a defined breeding period, the conception rate is appropriate. However, more recently, in seasonal and in year-round calving herds, the concept of conception risk has gained favour since it does not require a defined time period (Brownlie et al., 2014; Stevenson et al., 2014; Denis-Robichaud et al., 2016).

The submission rate (or risk) is the proportion of cows present in the herd at the start of mating that are bred (served or inseminated) in a defined time after breeding starts.

The InCalf programme reports this index as a 3-, 6- and 9-week rate after the planned start of mating date, which is the date occurring 282 days before the planned start of calving date (Anon, 2020). This index is influenced by the occurrence of oestrus, the detection of oestrus by farm workers in artificially inseminated cows and herds or bulls in naturally bred herds and then the insemination or breeding of the cows that are detected in oestrus. On high-performing farms, 3-week submission rates of 90% or more are achieved. High performance in this index is required for high in-calf rates to be achieved. The imperative for a high 3-week submission rate can lead to overzealous oestrus detection and the presentation of cows for insemination that are not in oestrus. Evidence for the occurrence of this practise is supported when conception rates are low in the first week of mating then noticeably improve.

The not-in-calf or empty rate is the percentage of the cows within a mating group that fail to become pregnant by the end of a mating period. This represents the proportion of the breeding group that fails to meet the desired outcome by the end of breeding.

Conception rates are the proportion of inseminations occurring in a defined period that result in conception. Since conception can only be measured after pregnancy has been diagnosed, the term pregnancy rate is sometimes used. However, there can be confusion in the use of the term pregnancy rate since, in North America, pregnancy rate can refer to the number of cows that are pregnant in a population of eligible cows that are beyond the voluntary waiting period for a defined time such as 21 days. In New Zealand, some herds can achieve conception rates to first service of 60% or greater, but rates above 53% are currently considered adequate. Conception rates are influenced by female (cow and heifer), male (bull and semen), management and environmental factors.

There are further indices of performance that assist in monitoring reproductive performance and help to pinpoint aspects of breeding that can be improved. Unlike the indices above, these performance measures are common to seasonal and year-round herds and are described below.

6 Production-related reproductive indices in year-round calving herds

In year-round calving herds, constraints on production differ from those in seasonal herds since a close tying of pasture availability and milk production does not exist. In these herds the main production-related indices are the calving interval, that is, the time between successive calving events, the culling rate and the proportion of a herd that calves in a year.

Calving intervals of 12 months in cows and up to 13 months in first lactation heifers have been shown to be optimal for milk production efficiency. Shorter calving intervals concentrate periods where cows produce milk at higher levels and with greater efficiency in the earlier parts of the lactation. The reason that heifers have longer recommended intervals is because their lactations persist at higher levels for longer. Shorter calving intervals are associated with a higher output of calves per cow.

The culling rate (for infertility) is another index related to production of animals available for sale – a farm output. The proportion of the herd that calves in a year is another production-related index in year-round calving herds that is not often monitored. A target for this index is greater than 110% due to some cows with intervals less than 12 months and to heifers or other new cows entering the herd only when they calve.

Reproductive intervals used as indices for heifers require analysis separate from those for cows because heifers are not regarded as a part of the milking herd until after their first calving. Heifer indices include several that are measured from birth. The time to puberty is the time from birth until the first oestrus, which should occur at 12–14 months. The time to first service should occur in the 13 to 15-month age range. The main production-related reproductive index for heifers is the time from birth to first calving. A suitable target for this index is from 22 to 24 months (Table 1).

There are analogous intervals for cows that are measured from the last previous calving date. The inter-calving (or calving) interval is ideally 12 months in year-round and seasonally calving herds. This interval index is calculated for cows only and excludes heifers since they do not have a previous calving record. It also excludes culled cows, and those that die or are otherwise disposed of, since no subsequent calving record is available for them. This is a production-related index but is limited because it is insensitive to recent change since it can only be calculated for those cows that do calve again.

The exclusion of cows that do not conceive or that are culled means that the inter-calving index is subject to biases due to omission of information from

Table 1 Reproductive intervals from birth for maiden heifers and calving for cows that are used to monitor performance

Interval name	Target	Apply to
Birth to first oestrus (Time to puberty)	12-14 months	Heifers
Birth to first service	13-15 months	Heifers
Birth to first calving	22-24 months	Heifers
Calving interval	12 months	Cows
Proportion of herd calving in a year	115%	Cows
Proportion of herd culled for reproduction	Less than 10%/year	Cows and heifers separately
Proportion of carry-over cows	Nil	Cows

poorly performing animals. For this reason, the inter-calving interval must be assessed in association with other indices to be validly interpreted. One is the culling rate while another is the number and proportion of cows that remain in the herd but are still not pregnant despite being well beyond the anticipated time of breeding, known as carry-over cows. Both increase the proportion of cows in the herd that is not pregnant and not milking and reduce the proportion that is diagnosed pregnant and milking.

The calving to conception interval is the mean time from calving to conception for a defined cohort of cows. The target for this interval is from 83 to 85 days in order to maintain a 12-month calving interval. Some call this the 'open interval' while others add on days to the last service or days to the current date to calculate the open interval for cows not yet diagnosed pregnant. The calving to conception and open intervals are highly correlated to the calving interval but considerably more current. How current the index is depends on the frequency of pregnancy diagnosis. The calving and open intervals are useful monitoring indices but are not useful in the causal diagnosis of problems because they are the result of male, female and management factors.

7 Indices used to diagnose causes of inadequate herd reproduction

The range of reproductive indices used to monitor performance is shown in Table 2. The calving to first service interval is the sum of the intervals to first service for all cows receiving first services, divided by the number of cows that receive a first service. This index helps to separate the causes of long calving intervals into late breeding due to either failure to breed or failure to conceive once bred. If the calving to first service interval is long, reasons for delayed breeding must be investigated. If calving to first service interval is short then long calving intervals can be due to a failure of cows to conceive to the early

Table 2 Reproductive indices used to monitor performance

Name of index	Target
Calving to first oestrus	45 days
Percentage in oestrus by 60 days after calving	>90%
Calving to first service	65 days
Percentage pregnant at pregnancy diagnosis	>85%
Oestrus detection rate (risk)	>85%
Ratio of single to double inter-oestrus intervals	>7:1
First service conception risk	45%
Services per conception	<1.7

breeding or a failure to detect and breed those cows that have not conceived and that return to oestrus after the first breeding. A target for the ideal calving to first service interval will depend on a herd's conception risk but will normally be in the range of 55-70 days. Underlying reasons for a prolonged interval from calving to first service can be anoestrus among the cows to be bred, a failure to detect oestrus by farm workers or a failure to breed cows at detected oestrus.

In order to sort out these potential causes, measures of the occurrence and detection of oestrus must be evaluated. The average calving to first oestrus interval should ideally be within 45 days of calving, but for this to be possible, all oestruses after calving must be recorded. The percentage of cows detected in oestrus by 60 days after calving should exceed 90% to indicate excellent return to oestrus by cows and oestrus detection by farm workers. Another index of the occurrence and detection of oestrus is the percentage of cows that are pregnant at PD in herds where all oestrus events are recorded. This measure reveals the adequacy of oestrus detection because if cows are bred and do not conceive then they should return to oestrus at a normal inter-oestrus interval and be detected. They should only be presented for PD if no oestrus has been observed after the last breeding. If they are not pregnant at PD but are otherwise normal this means that at least one and possibly two returns to oestrus have not been observed or noted. The target proportion pregnant at PD in herds where all observed oestrus events are recorded is greater than 85% if PD occurs at less than 45 days of gestation or over 95% if it occurs at over 45 days. The cycling status of cows can be checked at PD by confirming the presence of a CL on the ovaries of the non-pregnant cows examined.

The oestrus detection rate has also been described as a useful measure of the detection of oestrus. It is calculated as 21 divided by the average days between recorded oestruses times 100. The target for this index is over 85% but it can exceed 100% due to the false detection of oestruses, cows with short inter-oestrus intervals such as nymphomaniacs with cystic ovaries and, more commonly, due to the use of prostaglandins to induce oestrus which shortens ovarian cycles.

Another analytical technique that can be used to evaluate the efficiency of oestrus detection is the <u>ratio of single to double inter-oestrus intervals</u>. Since cows and heifers return to oestrus in a cyclical pattern, oestrus will occur in a peak at periods averaging 21 days after a previous oestrus. If an oestrus is not detected at around 21 days after a previous oestrus, the next occurrence is likely to be at around 42 days or two cycles later. This can be used to examine patterns of oestrus returns to evaluate if farm workers are missing too many occurrences of oestrus. The ratio of single to double inter-oestrus intervals should be from 7 to 1 or greater. This measure is very useful for indicating the efficiency of oestrus detection in large herds but, to be effective, all detected oestruses must be recorded and the use of agents to manipulate the length of oestrus cycles should be minimal.

The major influences on conception efficiency are sire, cow, management and environmental factors. The conception efficiency of cows is monitored as the <u>conception rate</u> now commonly named <u>conception risk</u> since in many cases no time element is included to make it a valid rate. Calculating conception risks to specific sires or inseminators may reveal individuals whose performance is noticeably below target or below that of peers. Enough first services within any groups examined are needed for results to be meaningful. Conception (pregnancy) risks also can be calculated for all services but these can be biased by cows that have many repeat services. Conception risks are often regarded as representing breeding efficiency because they are used to monitor the performance of artificial breeding centres, individual inseminators and individual sires. However, they are less economically important than calving intervals.

Since conception itself is difficult to determine, many proxies exist such as non-return rates and per service pregnancy risks. In view of this, it is critical to determine exactly how the conception index being assessed or monitored has been calculated. The <u>first service conception (by diagnosed pregnancy) risk</u> is the number of first services in a period resulting in a pregnancy divided by the number of first services conducted in the period, expressed as a percentage. It is a measure of the outcome of breeding in a group or herd and not of the performance of individual cows, since the result for individuals can only be 0% or 100%. The first service conception risk is used because it is relatively unbiased by cows that require repeated services due to problems. Frequently it is called the first service conception rate. This index can be focused to examine the outcome of breeding performed on specific age groups, groups mated by specific sires, groups housed together or bred by individual inseminators as the denominator in order to allow associations to be evaluated between these factors and conceptions achieved.

The 24-day non-return rate is the proportion of bred cows that do not return to oestrus by 24 days after breeding. It is sometimes used by farmers and insemination companies as an index of conception but is not an accurate measure of this outcome. A low value of this index does provide an early warning that upcoming conception rates are likely to be low. It overestimates conception risks by 10-25% due to factors such as cows not being re-bred, becoming anoestrus after breeding (phantom cows), subsequent oestrus not being detected or recorded, cows being sold or having died, cows being bred by herd bulls without this being recorded, cows inseminated by a different artificial breeding company and the occurrence of foetal death.

A further measure of breeding efficiency is the number of services per conception. This is the total number of services provided to a herd in a season or period, divided by the number of pregnancies achieved. A target for this index is less than 1.7. This index is useful in estimating the amount of semen required to enable a herd to be bred.

As mentioned above, 'pregnancy rate' in a North American context refers to the number of cows pregnant in a population of eligible cows where an eligible cow is beyond the 'voluntary waiting period' (bred and unbred cows) over a defined period that is usually 21 days, that is, 21-day pregnancy rate (Stevenson and Britt, 2017). This index provides a contemporary measure of breeding efficiency that accommodates high intervention breeding strategies used today. A measure of the occurrence of foetal loss should also be monitored and evaluated when abortions occur in dairy herds. This is discussed later in Section 14.

8 Monitoring bull breeding

Comparisons of results based on whether breeding has been by bulls in natural service or by artificial insemination (when both are being used in a herd) may highlight limits such as insemination technique or the ability of cows to conceive when their chances are not limited by the oestrus detection abilities on the farm. In seasonal pasture-based herds, bulls may be grazed together with the cows after the end of the period when artificial insemination is used. It is common that conception rates then improve. This may be due to the bulls' better performance due to their superior oestrus detection accuracy than that of farm workers, better timing of insemination and more frequent and higher sperm dosing. Since the timing of introduction of bulls to the herd commonly coincides with when cows achieve a positive energy balance after calving as production begins to decline, the improved fertility may also be partly due to this. Factors such as breed, age, weight, nutritional adequacy, time since calving, production level, health status and reproductive disease can all influence conception efficiency.

Where specific sires produce low conception rates, investigations or enquiries of performance in other herds using the same sire can be undertaken to further define a problem. The same can be done to examine the results obtained by inseminators. Checks can be conducted into semen quality, such as having the post-thaw motility of semen assessed by semen companies. The technique of inseminators can also be checked including their hygiene, thawing practice and semen placement, especially if they are farm employees, along with checks of their equipment.

The time of mating after calving and the time of mating within an oestrus both have major influences on the conception rates achieved. Early breeding that is used to align planned calving dates in late calving cows results in low conception rates. Breeding early or late within an oestrus also is associated with lower conception rates than when breeding occurs at optimal times, 12 to 18 hours after the commencement of oestrus. Conception rates early in the first 2 weeks of the breeding season in seasonally calving herds can be low due to a lack of specificity in oestrus detection when pre-breeding occurrences of oestrus are not recorded.

Inadequate conception rates by bulls when used in natural service require further investigation. A common management error is for farmers to have an inadequate number of bulls enter the herd to adequately cover the number of still non-pregnant cows after a period of artificial breeding because they over-estimate the conception efficiency that is achieved during the artificial breeding season. This leads to an underestimate of the number of bulls required to adequately breed the remaining non-pregnant cows. Although the number of cows that can be serviced by individual bulls varies, a ratio of 1 bull per 25 non-pregnant cows should be allowed to ensure breeding coverage.

Bull unsoundness can reduce reproductive efficiency and is frequently associated with feet and leg problems but also can be due to low libido, poor semen quality, injury or venereal disease. Bull breeding soundness examination conducted before the bull breeding period can help minimise breeding failure. In addition to locomotor system problems poor bull performance can also be due to low libido, low semen quality or venereal disease. Problems in cows generally do not cause infertility in the bull breeding cleanup period unless a venereal disease exists. This is now rare with the widespread use of artificial breeding because most cow problems have been resolved by the end of the artificial breeding period when bulls are placed with the milking herd.

9 Management of herd limits to reproduction: anoestrus

A major limit to herd reproductive efficiency is anoestrus because its occurrence delays breeding and extends calving intervals. This has been observed in studies conducted throughout the world in confined and pastured cattle with high and low production (Zemjanis et al., 1969; Fielden et al., 1973; McDougall et al., 1993; Mwaanga and Janowski, 2000; Macmillan, 2002; Ambrose, 2015). Anoestrus can delay conception with consequent losses in production and increased maintenance costs. To maintain a 365-day calving interval in cattle the average cow must conceive by the 83rd to 85th day after calving. Every oestrus missed after that means that her calving interval extends by about 3 weeks.

Anoestrus is not a disease. It is the manifestation of many conditions which affect the oestrous cycle. Clinically, it is presented as either an individual animal problem or revealed as a herd problem. Herd problems are most often identified after veterinary intervention in the form of an investigation of inadequate herd reproductive performance or by an analysis of performance as a part of an ongoing reproductive health programme.

In either case, the information presented to the herd veterinarian is dependent on the farm managers' abilities, observations and recording skills. When a farmer calls to have a cow examined because she has not been coming into oestrus, the farmer is the one making this assessment. Each case of anoestrus may be due to either a failure of signs of oestrus to occur or a failure of oestrus to be observed or recorded when it does occur.

Differentiation between these two possibilities is not always easy but knowledge of which problem exists is essential for appropriate intervention to manage the situation. Evaluation of records, thorough clinical evaluation and a thorough and detailed examination of oestrus detection and breeding practices are required to determine the likely causes of anoestrus and to allow appropriate intervention.

A common cause of anoestrus in cows at the planned start of mating in seasonally calving pasture-based herds is that insufficient time has elapsed for them to develop normal cycling patterns. Late calving cows are more prone to be anoestrus during the mating period and herds with spread calving patterns thus have more anoestrous cows.

Although ovarian follicles develop by 10 days after calving, ovulation may occur much later. In the United States, cows are reported as ovulating by 15 days after calving. In two Australian herds the author documented that ovulation occurred at approximately 24 days after calving. In New Zealand first ovulation occurred at 36 days in adult cows and at 52 days in 2- and 3-year-old cows. A more recent study documented ovulation at 44 days. The current herd calving pattern is the major determinant of the amount of anoestrus to be dealt with in seasonally calving dairy herds.

Factors that cause true anoestrus or a failure to cycle accounted for only 1/3 of the anoestrus observed in New Zealand dairy herds. About 2/3 of anoestrus (revealed when cows not seen in oestrus were examined a week before the planned start of mating) occurs in cows which are cycling normally (McDougall,

1994). Thus it is highly likely that herd management is at fault and that oestrus detection is the problem, rather than a failure of cows to cycle. Confirmation that the animal is cycling is based on a careful examination of the reproductive tract per rectum. In cycling cattle a CL will be palpable on an ovary for about 80% of the time. During stages when the CL is not palpable, follicular development and changes in tone of the tubular reproductive tract should indicate whether ovarian activity is present. If no evidence of uterine pathology exists cows are unlikely to have a retained CL and repeated examinations should confirm this. Although a cow is cycling, it still may have 'silent' oestrus. In practical terms checking the efficiency of oestrus detection procedures and 'predicting' when oestrus will occur will help clients to breed their animals. Further techniques to determine whether an animal is cycling use serial milk or serum progesterone assays but cost and availability are problems.

Cows presented for examination as being anoestrus may be cycling but not being observed. The many causes of real and apparent anoestrus in herds is called the 'anoestrus syndrome'. The prognosis, treatment and prevention of this problem relate to the cause, and so every effort must be made to determine what this is.

10 Clinical examination and treatment of anoestrous cows

Examination for anoestrus should be like any clinical examination of cows and include a thorough exploration of the cervix, uterus, ovaries and other reproductive organs. The examination is to determine if ovarian cycling is occurring and, if not, why not. Anoestrous cows can be divided into three groups based on the structures present on their ovaries. There may be *corpora luteum*, no structures or follicular structures.

10.1 Anoestrus with a corpus luteum present

Many cows presented because oestrus has not been observed are found to be pregnant, even though this finding is not consistent with the cows' records. All cows should be very carefully examined for pregnancy when they are presented as being anoestrous, because some treatments for anoestrus will terminate pregnancy. If there is doubt about the cow's status, re-examination later may clarify her status. Since many apparently anoestrous cows are found to be pregnant, routine examination of so-called anoestrous cows in herd health programmes should be delayed until the interval after a previous oestrus or service when a veterinarian can competently diagnose pregnancy. This helps prevent the inadvertent aborting of pregnant cows.

An apparently retained CL may be due to early embryonic death since, if the embryo dies after the 15th day of the oestrous cycle, the next oestrus is delayed. The small embryo may be aborted unseen or resorbed. This phenomenon can explain the higher incidence of anoestrus after service than anoestrus after oestrus. Progesterone studies indicate that this occurrence is relatively common, since many cows have elevated progesterone levels for 30-35 days after service, which then drop. The CL may also be retained when there is uterine distension or pathology, usually due to pyometra but also sometimes due to mucometra or foetal maceration or mummification.

Other causes of anoestrus with a retained CL are usually seen in barren heifers. Heifers with absence of the endometrial glands do not cycle but oestrus can be induced by removal of the CL manually. Presumably oestrus could be induced by causing luteolysis with prostaglandin, but these animals would probably remain sterile because they would be unlikely to be able to support pregnancy. Zemjanis et al. (1969) found that 2-3% of cows with diagnosed anoestrus had white heifer disease (segmental aplasia of the Mullerian ducts) whilst in New Zealand, the prevalence among cows culled after PD was only 0.09% (Lawton et al., 2000). The disease causes anoestrus if both horns, the uterine body, cervix or vagina are affected. If there is uterus unicornus, ovarian cycling and even pregnancy may occur. When a CL forms on the ovary adjacent to the aplastic horn, the CL may persist indefinitely causing anoestrus. In cases of congenital defects, the rational treatment is to cull affected heifers.

Treatment of anoestrus cows with a CL present will vary depending on other findings. Pyometra, foetal maceration or mummification and mucometra are best treated by prostaglandin-induced luteolysis, which should help to evacuate the uterus and start normal cycling. Other possibilities for treatment include ergonovine, oestradiol and oxytocin. Supportive treatment with antibiotic infusion to the uterus may be given for pyometra. The treatments commonly used are 1-5 g of oxytetracycline-active ingredient or 2 million units of penicillin with or without 2 g of streptomycin. These are generally infused in 20-100 mL of water.

Cows with anoestrus due to suboestrus, weak or silent oestrus also may have a CL palpable on their ovaries. Oestrus may not be observed in these cows for up to 120 days or more after calving when competent herdsmen observe them. On repeated rectal examinations, cows show evidence of cycling with corpora lutea changing sides. Metoestral bleeding or mucus discharge associated with uterine tone may also be found, and some cows may have ovulation depressions (OVDs) on their ovaries. About 85% of normal cycling cows should have readily palpable CLs on their ovaries.

Suboestrus or 'Silent heat' indicates that an animal goes through an ovarian cycle and ovulates but does not show oestrus. Ovulation, without overt signs of oestrus, can be recognised by examination per rectum. These ovulations may be 'fertile cycles' if artificial insemination is used to breed the animal at an appropriate time. Suboestrus occurs, but weak oestrus and inadequate detection are almost indistinguishable. The reported incidence of suboestrus ranges from 18% to 30%. Silent oestrus is common in early ovulations after calving. Data from many studies indicate that 60% of first ovulations after calving are silent. The incidence declines with subsequent cycles until after 60 days post-partum, the incidence is only 6-10%.

Silent oestrus is a convenient diagnosis to make to a farmer when the alternative is to diagnose inadequate oestrus detection, but that does not help the farmer if inadequate detection is indeed the cause of the problem. In herds where oestrus detection is excellent, 21-day submission rates may exceed 95% and mean intervals from calving to first oestrus may be 23-25 days, even in large herds.

The author has an adage called the 'law of silent oestrus' that states 'The incidence of silent oestrus is inversely proportional to the farmer's effort and skill at oestrus detection.'

Several factors influence the expression of oestrus including breed, line, presence of metritis, high production, milking frequency, suckling, high concentrate feeding and lameness. Some of these factors can be eliminated or dealt with while others must be endured. The major influence on the contribution of weak and silent oestrus to an anoestrus problem is the amount of effort and expertise applied in oestrus detection.

10.2 Anoestrus with inactive ovaries

Anoestrous cows or heifers with inactive ovaries include those with congenital abnormalities of the genitalia, for example, ovarian hypoplasia and freemartinism and more commonly, those undergoing a nutritional deficiency (stress). Ovarian hypoplasia is reported in the Shorthorn and Swedish Highland breeds of cattle. In them it is genetically transmitted as an autosomal recessive gene with incomplete penetrance. It can also occur sporadically in other breeds. It can be partial or complete, unilateral or bilateral. Anoestrus only occurs when the disease is bilateral and complete. Unilateral and incomplete hypoplasia does not cause sterility so there is a possibility of the disease being spread in the cattle population if carriers are not recognised. The ovaries in affected heifers and cows are thin, firm, narrow structures in the broad ligament. If bilateral, the genitalia are infantile, and the conformation of the heifers is steer-like. Partial hypoplasia involves the medial part of the ovary and the lateral pole may contain normal structures in externally normal-looking cows. Affected animals should be slaughtered.

Freemartinism is the most common congenital defect causing anoestrus. Freemartins comprised 9.5% of clinically diagnosed causes of anoestrus in Minnesota in the study by Zemjanis et al. (1969). The disease is restricted to females born co-twins to bulls and 90% of these are affected. About 8-9% are fertile. The Holstein breed has the highest incidence due to the highest rate of multiple di-zygotic births. The condition is due to a humoral or cellular factor passing from the male to the female twin through anastomosing placental vessels before the stage of sexual differentiation. Freemartins have small undifferentiated ovaries and small and undifferentiated vagina and uterus. A normal vulva and vestibule are present. The condition is diagnosed when there is a history of being a co-twin to a bull and a test tube cannot be inserted more than about 7.5 cm into the genital tract of a freemartin whereas in a normal female the tube will pass from 12 cm to 15 cm into the vagina. There is XX and XY chimerism of leucocyte cultures observable on cytogenetic examination. Affected animals are sterile and therefore culling is the recommended outcome for these animals.

True and pseudo-hermaphrodite conditions that are associated with anoestrus occur in cattle. True hermaphrodites and male pseudo-hermaphrodites may look like females. They are frequently first noticed because of abnormalities in genitalia or male-like body conformation and behaviour. Various types of chimerism may be present in these animals.

The most common cause of anoestrus is due to underfeeding and stress which occur in heifers and cows. In heifers there is an inverse relationship between energy intake during rearing and age at puberty. Many studies in cows show that the level of feeding before calving has a marked influence on the interval until resumption of oestrus after calving. The level of feeding after calving also has an influence on this interval, but not as great as the feeding level before calving which influences the body condition at calving. High nutritional demands for production compete for resources with reproductive requirements and cause changes in body composition associated with reproductive outcomes (Chagas et al., 2007).

Studies of dairy cattle in Australia and New Zealand have found a negative correlation between body condition at calving and the interval to oestrus. Experimental work shows that cows with high body condition at calving are better able to cope with nutritional stresses after calving. Hypoglycaemia has been related to a delay in the onset of oestrus after calving, both in field situations and when hypoglycaemia has been experimentally induced with insulin. Low planes of nutrition, and even short-term fasting, decrease the ovarian response to gonadotrophins which are 'injected' and a similar mechanism may apply in undernourished cows with regard to endogenous gonadotrophins.

The diagnosis of anoestrus due to energy deficiency can frequently be made by assessing body condition and measuring body weight. If cows have low weights and low condition, then low energy intake is the likely cause of their problems. A low blood glucose level in the group of animals affected and high beta-hydroxybuterate level confirm this. Where pasture availability is monitored as a part of nutritional services, a lack of enough feed and an inadequate intake to meet requirements can be documented. An assessment of the diet or pasture intake will reveal undernutrition, when compared to accepted feeding standards.

Treatment is by providing adequate nutrition which should also prevent the problem. In some cases in New Zealand, the problem has been addressed by reducing the nutritional demands on cows. Applying once per day milking until cows are mated can do this by reducing energy requirements. When done in association with supplementary feeding of concentrates, anecdotal reports are that oestrus cycling and conception improved in herds that adopted this strategy in the face of severe feed shortages. Protein deficiency in feed may cause anoestrus but this is due to its effect in depressing appetite and limiting energy utilisation.

Phosphorus deficiency can cause anoestrus through primary or secondary effects. There is a strong association of phosphorus deficiency with poor conception rates and anoestrus that may be due to low protein levels in phosphorus-deficient feeds, low vitamin A levels often associated with phosphorus-deficient feeds, and almost certainly the depressed appetite and wasting which accompanies phosphorus deficiency. Diagnosis can be made in the presence of mean herd blood phosphorus levels of less than 4 mg per 100 mL or when signs of clinical phosphorus deficiency are present. Phosphorus intakes can also be compared with standards. Controlled response trials can confirm a diagnosis of any nutritional deficiency.

Calcium/phosphorus ratio may also be important, and a study showed that anoestrus occurred if there was imbalance in calcium/phosphorus ratios with inadequate vitamin D. It appears that high calcium/low phosphorus ratios inhibit the expression of oestrus and the occurrence of ovulation, particularly if vitamin D is inadequate. Manganese deficiency has also been associated with anoestrus and suboestrus, but always accompanied by the birth of weak calves and calves with arthrogryposis. High dietary calcium may predispose to manganese deficiency.

Copper deficiency has been associated with anoestrus and infertility but the evidence for a direct effect is poor. Anoestrus due to copper deficiency seems only to occur when the signs of copper deficiency are obvious and when anaemia is present. Anaemia with <8 g haemoglobin/100 mL blood is likely to cause anoestrus. The anaemia, weight loss and poor growth, typical of copper deficiency, may also be intermediate causes in the anoestrus seen in copper deficiency. Diagnosis can be made based on signs of copper deficiency and blood levels of <0.07 mg/100 mL and feed levels of <20 ppm dry matter. Supplementation can be used for prevention and control with 1-2 g of $CuSo_4$ daily. Copper glycinate injections of 150 mg for 2-3 times per year were very popular in Australia. Copper wire boluses are popular in New Zealand but are expensive. Inclusion of $CuSO_4$ in topdressing applications is cheap and relatively effective.

Any disease which causes debility or anorexia and loss of body weight can result in a failure of oestrus. The pathogenesis may be due to the loss of body weight, low blood glucose levels, weight loss or the anaemia commonly associated with debilitating diseases.

10.3 Anoestrus with follicular or cystic structures

At examination, anoestrous cows may also have follicular structures or cysts present on their ovaries. If cows are found to have follicles on their ovaries at examination for anoestrus, they may be cycling normally, and may not have been observed by the farmer. Ovarian cysts were found to be the most common cause of true or 'organic' anoestrus in the Minnesota study of Zemjanis et al. (1969) that is, non-cycling anoestrus. At least 50% of cows with follicular cystic ovaries are anoestrous. Luteal cysts are usually associated with anoestrus (i.e. luteinised anovulating follicles). Cows with cystic CL which develop after ovulation cycle normally and are not anoestrous for prolonged periods.

There may be one or more cystic follicles on one or both ovaries. They are generally soft, spherical, fluctuating structures which feel like they project from the ovarian surface. They generally exceed 2 cm in diameter. Cysts feel thicker walled than normal follicles. Some are easily ruptured, but luteal cysts may be very difficult to rupture. The mucus from cystic cows is scantier and less clear than from oestrous cows. Some cows show elevation of the tail head and relaxation of the sacro-sciatic ligaments. The uterus may become thick-walled, flaccid and the horns feel shortened. Several treatments are available including gonadotrophin-releasing hormone being the favoured treatment, pituitary gonadotrophins having a luteotrophic effect, human chorionic gonadotrophin for follicular cysts and prostaglandins for cows with luteal cysts. Spontaneous recovery may occur but is less likely with increasing time from calving. Brito and Palmer (2004) extensively reviewed incidence, pathogenesis, diagnosis and treatment of cystic ovarian disease.

10.4 Treatment of the anoestrus syndrome

Pregnancy and other physiological conditions must be ruled out before any form of treatment is instituted. Where pathology of the gonads or of the tubular genital tract or intercurrent disease exists, each case must be dealt with on its merits. In many cases improved husbandry will be required. Fortunately, most cases of anoestrus in cattle are temporary and, given time, most affected cows cycle and become pregnant. The problem is in the delays that occur especially under conditions of seasonal breeding.

In managing pre-service anoestrus, the first step is to define the problem in the herd in question. This means careful examination of herd calving and breeding records and clinical examination of problem animals. Pregnancy must be protected if present and obvious 'disease' recognised. Excluding diseased and pregnant cases, the animals being examined fall into two groups:

- those with active ovaries; and
- those with inactive ovaries.

Where ovarian activity exists, this generally means the presence of a CL with no pathology of the genital tract, or cows that are approaching, in, or just going out of, oestrus. The prognosis is thus favourable and nothing need to be done except make the client aware that these cows are cycling; one should endeavour to predict when each will be in oestrus. Once alerted, the client can take more care with detection procedures. If many of these animals are noted in a herd, oestrus detection processes need checking.

Injectable prostaglandins or analogues, progesterone treatments, or Ovsynch programmes are used where circumstances warrant. With these techniques, oestrus occurrence is synchronised and prediction of oestrus and ovulation is controlled such that appointment breeding can be used. This eliminates the need for oestrus detection and fertility may be enhanced in some cases. Cost/benefit must be considered before embarking on such programmes.

Where ovarian inactivity exists, preferential feeding of potential or actual problem animals is required. Factors other than nutrition do influence return to post-partum cyclical activity (e.g. the frequency of suckling or milking, sequelae to difficult birth, etc.), but feed intake and treatment of current disease are two factors which may be managerially controlled. Short-term alteration of feed input has remarkably little effect on returns to cyclical activity if the animals concerned are in a 'reproductively unfit' state to recommence the reproductive process again. Cows and heifers calving in poor condition generally divert nutrients to meet the needs of milk production and growth rather than reproduction. Even if ad libitum feed is made available, there is likely to be a considerable lag before a return to cyclical activity occurs. Heifer and cows calving in adequate body condition have the potential to 'survive' this initial lactational stress period and will respond by cycling much sooner unless very severe nutritional deprivation occurs. The establishment of minimum target weights (and/or condition scores) for animals at breeding and calving is a sound practice. Management practices should be examined to ensure that targets can be achieved including the grazing or feeding system, fertiliser practices, 'drying off' dates and animal health interventions.

Mating of heifers 10-14 days before breeding commences in the milking herd is recommended to increase submission rates for primiparous animals. Pre-service anoestrus with inactive ovaries has proven to be significant problem with first calf heifers. By adopting this early mating procedure, the mean calving date for the heifers is usually about 16 days ahead of the mean calving date for the main breeding herd. Their requirement for a relatively long post-partum period before resuming ovarian activity is thus compensated for and they can be expected to have a similar second calving pattern to the rest of the herd. Where satisfactory submission rates with these 2-year old animals are being achieved early breeding of the heifers is unnecessary.

11 Improving oestrus detection

Failure to observe oestrus is due to many factors. There is a variation in the length and intensity of oestrus. Studies show that there are periods of intense oestrus activity interspersed by quiet periods. A failure to observe cows at appropriate times may influence oestrus detection and some cows with short duration of oestrus will be missed unless observation coincides with their display of oestrus. Studies show that cows in barns have oestrus which lasts only for 8 hours in contrast to the commonly cited oestrus length of 16–18 hours for cows at pasture. There can be considerable variation between individuals in the duration and intensity of the display of oestrus. Thus while 18 hours is an average for cows to show standing oestrus the range may vary from 4 hours to 30 hours. Since the intensity of the signs shown also varies, it is understandable, particularly in large herds, why the signs may be missed.

Inadequate observation and ignorance of the behavioural and physical signs of oestrus are the major causes of apparent anoestrus. Many farms do not have set periods for oestrus detection, while others only observe once per day, when three times for daily observation is required for optimal oestrus detection. When cows are at pasture, observation for oestrus activity needs to be scheduled as a farm task rather than something which occurs as other farm tasks (such as collecting the cows for milking) are done.

Increased herd sizes make detection more difficult. The facilities needed for larger herds make observation more difficult and the observation time per cow is reduced, since labour availability generally is not increased in proportion to herd size. The importance of heat detection, evaluation of heat detection efficiency, causes of inadequate detection and methods to improve detection, including heat detection aids, were reviewed by Williamson (2005).

Where inadequate oestrus detection is limiting herd reproductive performance the major step in improving the situation is to convince farmers that oestrus detection is an important farm task. To help do this, the oestrus detection efficiency of the farm can be demonstrated by analysing records. The use of aids, like pressure detectors, rubbing detectors or chin-ball markers on teaser bulls for a short period, may show the farmer that many oestruses are being missed. The economic importance of oestrus detection should be stressed to motivate farmers to improve it. False positive detection can be documented by taking serum or milk samples at observed oestrus and analysing for progesterone levels. The presence of a high frequency of progesterone above baseline levels indicates false positive detection.

To achieve improvements the farmer should be encouraged to set aside time to observe cows for oestrus. The time of observation is important. It should be when cows are not being milked and when they are not eating or being moved. A period when the majority of the herd is recumbent and ruminating is the best time. Three periods of 20–30 minutes at times when the herd settles for rumination should be enough to detect most oestrus events. Where cows are restrained in barns the most important step is to let them out for observation. If they are to be fed, then observation should be long enough after they are fed for most of the herd to have settled down and be ruminating. This also applies to situations where bunk feeding occurs in otherwise pastured or lot-housed cattle.

In order to achieve adequate oestrus detection, farmers and farm workers must be educated in the primary or definitive signs of oestrus and the secondary or suggestive signs. Primary signs of oestrus are standing to be mounted and mounting the head of another cow. Recognition of secondary signs greatly assists in detecting oestrus. These include restlessness, mooing and bellowing, increased activity, and standing when most of the herd is lying down. Cows in and near oestrus display seeking activity that includes nudging, sniffing and rubbing their chins on the rear end of other cows. Cows in or near oestrus have wet, pink and swollen vulvae, with long clear mucus discharges. They show a decrease in appetite and milk production. The most visible sign is that they mount other cows. The recording of all observed oestruses allows the prediction of subsequent ones and ensures that non-cycling cows with pathology or those needing therapy to initiate cycling can be identified. Oestrus prediction charts and computer-generated lists are available to aid in predicting upcoming oestrus events. Metoestral bleeding can be used to alert the client to watch for oestrus in 16–22 days.

Oestrus detection aids are invaluable in assisting accurate oestrus detection to occur. Several types of detection patches are available including pressure-sensitive detectors that are glued to the rumps of cows and undergo a colour change when cows are mounted. In a study conducted in Australia, herdsmen detected 56% of oestrus in pastured dairy cattle which did not have detectors fitted and in 90% of cows which did (Williamson et al., 1972). The use of detectors can be highly profitable where oestrus detection is improved.

The use of 'tail paint' or chalk applied over the sacral and first few coccygeal spines of cows acts as a cheap and effective alternative to oestrus detection patches. The paint becomes rubbed and broken when cows are mounted. The paint needs to be 'touched up' or freshened about once per week when latex paints are used and each 2–3 weeks when enamels are used.

Teaser animals can be used to aid oestrus detection and they are especially useful if fitted with Chin-ball harnesses. Teaser animals may be bulls with Chin-ball harnesses and with vasectomies, epididymectomies or deviated penises. They may also be testosterone-primed steers or cows or nymphomaniac cows. Injecting cows or steers with intra-muscular testosterone propionate, 950 mg every 2 days for 20 days, then 500 mg every 2-3 weeks can create teaser animals. These teaser animals mark oestrous cows and help identify them. Cows with chronically cystic ovaries can make effective teaser animals.

Veterinarians may predict oestrus when operating herd health programmes by assessing the state of the genital tract at palpation or scanning. The accuracy of prediction is increased greatly if the animal has a mature mid-cycle CL present on its ovary. This can be made to undergo luteolysis by injecting prostaglandin. Oestrus then usually occurs in 2-4 days and animals can be bred by appointment A.I. at 72 and 96 hours or 80 hours after treatment.

Many further aids to oestrus detection are available including electronic means of reading sensor strips affixed to cows, motion sensors mounted as pedometers or necklets, accelerometers that measure jumping activity, electronic pressure detectors that can link to computers in systems that may alert farmers to the occurrence of oestrus. Some further aids to oestrus detection rely on physiological aids to oestrus detection such as vaginal mucus conductivity probes, progesterone analysis kits to measure low progesterone levels in plasma or serum associated with oestrus. Temperature probes and systems to measure decreases in feed intake and milk production have also been coupled with other means to help detect the occurrence of oestrus in cows.

12 Controlled breeding programmes for oestrus synchronisation

There are three main types of oestrus synchronisation programmes used in the management of cattle breeding.

- Those based on extension of the luteal phase of oestrous cycles using exogenous progesterone or progestogens.
- Those based on luteolysis using prostaglandins to shorten the luteal phase of the cycle.
- Those based on follicular wave control by luteinisation and synchronisation of follicular wave development and LH release with gonadotrophin releasing hormone (GnRH) to ensure follicles are young at ovulation.

Many variations on these programmes exist and they may be supplemented using additional treatments, and aspects of the programmes may be mixed together.

When synchronisation programmes are being introduced to farms, adequate management is required for programmes to succeed. Adequate management includes:

- Clarity on the purpose of synchronisation – that is, is it to allow AB, minimise oestrus detection, help treat anoestrus or reduce calving spread?
- Careful planning and documentation of procedures, since programmed steps are required at the appropriate times when people and facilities must be available.
- Complete records to select animals for treatment and conduct the programme.
- High semen quality and serving capacity for natural sires with sufficient inseminators and/or bulls available.
- Experienced and capable inseminators if many animals are to be bred.
- Adequate and efficient facilities.

Animals selected for controlled breeding programmes should be free of disease and abnormality, adequately grown, well nourished, recovered from parturition and preferably cycling. This can be determined by detecting presence of a CL or by oestrus detection prior to implementing the programme. They also should be confirmed as not pregnant.

12.1 Progesterone-based programmes to induce and synchronise oestrus

Progestogen or progesterone-based devices are available for controlled breeding in cattle, including intravaginal devices that contain progesterone in silastic coatings or pods and silastic subcutaneous implants. These devices provide a prolonged, measured release of progesterone or a progestogen to simulate the influence of a CL and thereby prolong the luteal phase of the oestrus cycle so that the time of onset of oestrus can be controlled by acute withdrawal of the progesterone or progestogen.

Orally active progestogens were researched to provide a mechanism for controlled breeding, but the programmes of administration proved difficult for farmers to follow, and their use was associated with a reduction of conception rates.

These devices may be used for varying times in cows and result in relatively high synchrony of oestrus and ovulation if the treatment period with progesterone is sufficiently long. However prolonged progesterone exposure lowers the fertility of cows at the controlled oestrus. This is because progesterone or progestogens suppress LH pulse frequency and this causes suppression of dominant follicle growth. Progesterone does not, however, suppress FSH secretion. Progesterone treatment slows the growth of the dominant follicle and will prevent ovulation but does not stop or synchronise regular periodic FSH surges and emergence of follicular waves. Lowered fertility after prolonged progestogen treatment (>14 days) observed in early studies has been attributed to aging of the oocyte within the long-lived oversized follicles. Oestrus tends to occur on the 2nd and 3rd days after removal of the source of progesterone. Combined progesterone and prostaglandin treatments produce synchronisation of oestrus with a high degree of synchrony and high fertility. A progesterone device is used for 7-9 days and a prostaglandin treatment is administered at the time of removal of the progesterone source. Half-doses of prostaglandin are effective in this programme.

Anoestrus, low submission rates and poor breeding outcomes in North America have led to a system that requires hormonally controlled breeding to enable conceptions to occur. Also in seasonal pasture-based herds, controlled breeding is sometimes required to achieve compact calving and schemes exist to enhance the efficiency of breeding through hormonal intervention. Developments have occurred in the use of progestins and prostaglandins as synchronising agents, sometimes supplemented by other hormones such as oestrogens (where legally allowed) and hormones to stimulate follicular development as well as the use of GnRH plus prostaglandin and steroids as a means of controlling follicular wave development and inducing ovulation.

Induction of oestrus cycling as well as synchronisation of oestrus can be achieved through progesterone treatment, often combined with other hormones. If heifers are under-grown and undernourished, such progesterone treatment is ineffective. In such cases correction and prevention of further recurrences should be directed at removing underlying causes.

If farmers wish to treat non-cycling anoestrous cows and heifers to induce ovarian cycling, they may treat them with progesterone devices for 7 days and then remove the devices and inject with 400 IU of PMSG. Cows should then be watched closely for return to oestrus and inseminated on detection of oestrus. If no oestrus is observed within 3 weeks of device insertion, cows can be re-examined and a second treatment regime started if no ovarian activity is detected, or they can be treated with prostaglandin if a CL is palpated on an ovary. A recommended programme for inducing oestrus in non-cycling cows and heifers in New Zealand involves inserting a CIDR-B on day 0, removing the CIDR on day 5 and then injecting with PMSG at the time of CIDR removal. The injection of GnRH at or after CIDR removal causes an LH surge and more synchronous ovulation. Conception rates and pregnancy rates vary but are acceptable and similar to those in control groups. Double insemination at 48-60 hours after removal has produced pregnancy rates up to 75%.

In evaluating the efficacy of induction of oestrus with progesterone devices, results are measured in terms of conceptions and not just submission rates since there is evidence that false positive detections increase. The advantage of induced oestrus is partly its earlier occurrence. In evaluating the economic advantage of induction of oestrus, controls are essential since it has been observed in one study that 41% of control cows showed oestrus within 7 days of treatment of the treatment group cows. In another study, 60% of control cows showed oestrus within 21 days of treatment while an average of about 65% of treated cows showed oestrus within 7 days of treatment.

12.2 Prostaglandin-based synchronisation programmes

Several approaches to synchronisation of oestrus with prostaglandin are used. The product used is largely immaterial since available prostaglandin products show similar efficacy in causing luteolysis at recommended dose rates. Prostaglandin causes luteolysis of a mature CL3 on an ovary. Luteolysis only occurs if a mature CL is present. Corpora haemorrhagicum are refractory to prostaglandins until about the 6th day of the cycle. Studies show that maximal response to treatment in lactating dairy cows does not occur until day 8 of the cycle when over 90% undergo luteolysis. Cows respond well to prostaglandin-induced luteolysis between days 8 and 18 of the cycle. Following day 18 the regressing CL is not responsive to injected prostaglandin, but is regressing anyway due to the endogenous hormone.

Prostaglandin $F_{2\alpha}$ (25 mg) or cloprostenol (500 µg) given intramuscularly causes luteolysis and subsequent oestrus. Schemes have been devised to treat groups of animals to synchronise oestrus so that they may then be bred by appointment insemination, thus eliminating the need for oestrus detection. Schemes involving palpation or oestrus detection followed by single injections of these products or double injection programmes 11 days apart and then followed by insemination at 80 hours after the second treatment or twice at 72 and 96 hours have achieved fertility comparable to or better than that achieved in control animals. Controlled oestrus has the potential to reduce the spread of breeding and to achieve early compact calving with reduced labour required for oestrus detection, which results in longer lactations for the treated animals.

A double injection prostaglandin programme allows a high degree of synchrony of oestrus and ovulation to be achieved in randomly cycling animals. Suitable cycling cattle are injected on the first day of the programme. Those in dioestrus will undergo luteolysis and show oestrus, ovulate and form a new CL. Those in metoestrus will continue to develop a CL. Those in proestrus will continue to progress towards oestrus, in synchrony with the dioestrous cows that were injected with prostaglandin. The cows or heifers are then treated again 11 days after the first treatment when all would be expected to have a mature CL3 on their ovaries. Animals can be inseminated at detected oestrus, at 72 and 96 hours after the second injection or once at 72 hours (60 hours for heifers and dry cows). Fertility is highest when oestrus is detected.

In a single/double injection programme, suitable cycling cattle are injected with prostaglandin on day 1 of the programme and observed for oestrus over the next 5-7 days. Those detected in oestrus are inseminated and the remainder receive a second injection 11 days after the first. Animals receiving the second injection may be inseminated by set time breeding as for the double injection programme or after detection of oestrus. In a single injection programme, cows and heifers are palpated and if a mature CL is detected, they are treated with a dose of prostaglandin or analogue. Breeding is as for the previous two treatment programmes.

In the detection plus injection or 'Why Wait program', oestrus is observed carefully (using detectors or tail paint as aids) from 11 days before the planned start of mating (PSM). Cows in oestrus from 11 days to 6 days before the planned start of mating are injected with prostaglandin at the planned start of mating. They are then bred at detected oestrus, which is likely to occur 3-4 days later. Those in oestrus from 5 days to 1 day before the planned start of mating are injected with prostaglandin 6 days into mating and are inseminated at detected oestrus, from 9 days to 10 days into the mating season. Use of this and similar systems has resulted in mean intervals from planned start of mating to first insemination of 7.2 days with a 91% 21-day submission rate in New Zealand herds.

The detection, insemination, injection programme starts with oestrus detection for 5-6 days and cows are inseminated on detection. If over 25% of the group is inseminated, it indicates a high level of cyclicity in the group and the remainder are injected on day 6 with prostaglandin. Insemination then continues for up to a week after injection. This method is most suited to groups of heifers for AI since it allows their cycling status to be evaluated.

12.3 Gonadotrophin-prostaglandin programmes

The OvSynch programme uses GnRH followed by a prostaglandin $F_2\alpha$ injection 7 days later and a second GnRH treatment 24 or 48 hours after the prostaglandin treatment. This regime of treatment has resulted in synchronised oestrus in both cycling and non-cycling cows. After the second injection of GnRH the cows are bred without regard to oestrus behaviour with the optimum time after GnRH being reported as 16 hours (Pursley et al., 1995). There is considerable variability in response to this treatment programme. Results appear to be influenced by the cycling status of cows, their age (heifers respond poorly) and the precise timing of treatments and insemination. A meta-analysis found that the fertility of cows bred is around 40% conception rates where this regime has been implemented. If this is normal for an area this may be acceptable, but it may represent a reduction in fertility in some environments. The cost and multiple treatments involved in the Ovsynch programme may limit its use in some countries (Rabiee et al., 2005).

12.4 Combination programmes

Programmes have become very mixed so that there are various combinations and joint delivery of aspects of the programmes mentioned above. Detailed programmes are too numerous to review here, but include combinations of progesterone, oestradiol or GnRH or HCG and cloprostenol (Gyawu et al., 1991) and combinations of GnRH-prostaglandin programmes with progesterone intravaginal inserts (McDougall and Compton, 2008; Stevenson and Britt, 2017).

12.5 Pre-synchronisation and resynchronisation programmes

The various programmes for induction and synchronisation of oestrus followed by timed artificial insemination have been extensively reviewed recently (Stevenson and Britt, 2017; Fricke, 2019). They also describe resynchronisation programmes that may start 7 days before or at the time of a negative PD of cows that have previously been bred by appointment insemination, have not returned to oestrus but are not pregnant (elsewhere called phantom cows). These programmes start with an injection of GnRH or HCG 5 days or 7 days before or on the day of negative PD as the start of the next synchronisation cycle.

13 The role of nutrition in limiting and optimising reproduction

Nutritional limits to reproduction are universal in dairy farming. They may involve deficiencies or excesses of nutrients, usually the former, and other ingested matter such as toxins present within feeds. The impact of nutrients on fertility is manifest in many reproductive indicators such as delayed onset of puberty, anoestrus, failure to conceive and early embryonic death. Inadequate levels of specific nutrients in the diet and imbalances of some nutrients may inhibit the occurrence and display of oestrus, cause low conception rates, abortions, dystocia and the birth of weak offspring.

13.1 Plant and fungal toxins

Phyto-oestrogens and fungal oestrogens have been associated with the development of cystic ovaries and uterine pathologies that limit conception rates. Phyto-oestrogens depress fertility, possibly due to the cystic endometrial disease that they cause and may cause cystic ovaries. Sources of phyto-estrogens are some species of clovers, alfalfa and moulds that produce zearalenone.

Plant contaminants of pasture or trees consumed, specifically the needles of *Cupressus macrocarpa* and *Pinus ponderosa,* contain isocupressic acid which is an abortifacient in cattle if enough toxin is ingested.

13.2 Energy intake and balance

A primary aspect of nutrition with impacts on reproduction is the total feeding amount. Studies have shown that heifers fed inadequate rations and those fed plentiful rations achieve puberty late and early, respectively, but at about the same body size. More recently, it has been shown that increasing the feed level does increase the weight at which puberty occurs. Once oestrous cycling occurs, generally it continues if the rate of feeding is maintained. Oestrous cycling can be stopped by severe nutritional stress and cycling resumes when nutritional levels improve and body condition is restored.

Nutritional anoestrus of young heifers is exhibited as small, flattened, smooth, even-shaped ovaries and an underdeveloped tubular tract if the heifer has not reached puberty. In older heifers the ovaries are more likely to be 'knobbly' as follicular activity starts but with no readily discernible structures. Silent oestrus commonly occurs in the first one or two cycles of heifers reaching puberty. Heifers also weigh less than the weight expected at puberty which in New Zealand is:

- Holstein/Friesians   240-290 kg.
- Jerseys   163-204 kg.
- Shorthorns   240 kg.

Lactating dairy cows have high feed requirements because of the high milk outputs that they achieve and the need to provide nutrition to support this milk production in addition to growth and maintenance requirements. Closely allied with this is the energy required by dairy cows. Energy deficiency can be caused by a lack of feed, inadequate feed intake, poor-quality feed or a deficiency of intake relative to production. In pasture-based systems, cows have poor fertility when feeding is inadequate. Cows also have low body condition scores. Adequate feed intakes and balance underlie adequate reproduction.

13.3 Protein

Protein intakes can also be inadequate under some circumstances such as when cows graze dry-standing feed in dry summer and autumn conditions. Protein can also be excessive in some springs when pasture growth is high and pastures are lush. Under conditions of fully feeding cows or zero grazing, protein intakes are frequently controlled because feeds used as protein sources are generally expensive and so are limited to levels unlikely to cause infertility due to excesses in formulated rations.

13.4 Macrominerals

Some mineral deficiencies are known to be causal factors for retained placenta and metritis as well as reduced bull fertility. Calcium and phosphorus macro-minerals impact many aspects of reproduction. Older studies showed that significant effects of phosphorus deficiency are depression of appetite leading to underfeeding and energy deficiency, anoestrus, suboestrus, irregular oestrus cycles and low conception rates. Calcium to phosphorus ratios have also been implicated in causing infertility when they are well away from 1:1. These effects are exacerbated by low vitamin D. Interactions of calcium and phosphorus and manganese in causing infertility have also been reported in England. These may reflect a previously unrecognised dietary cation-anion difference effect. Calcium deficiency and hypocalcaemia may also reduce fertility due to decreases in muscle tone of the uterus with subsequent dystocias, retained foetal membranes and subsequent metritis or endometritis. Several studies also suggest that an excess dietary cation-anion difference reduces conception rates.

13.5 Trace elements

Trace elements also have an impact on fertility. Iodine deficiency has been reported to cause anoestrus, suboestrus, low conception rates, abortion, retained placenta and the birth of dead or weak calves that have arthrogryposis. In addition to its interaction with the calcium to phosphorus ratio, manganese deficiency can cause anoestrus, suboestrus and reduced conception rates. When manganese deficiency is associated with these reproductive signs, it is also associated with the birth of dead or weak calves that have arthrogryposis. This has been reported in New Zealand associated with apple pomice feeding. Many reports associate copper deficiency (direct or secondary) with infertility exhibited as anoestrus and poor conception rates. These effects may be secondary to reduced feed intake, anaemia and reduced body condition that are associated with copper deficiency. It is not likely that copper deficiency will be the cause of poor reproductive performance in the absence of other signs of deficiency. Cobalt deficiency can cause infertility secondary to inappetence, cachexia and debility associated with concurrent anaemia. The reproductive signs are anoestrus, suboestrus and reduced conception rates. Selenium deficiency is a cause of retained placenta, poor uterine involution and metritis. In New Zealand studies, supplementation with selenium failed to improve fertility despite improvements in growth and milk production occurring. Zinc deficiency is a cause of testicular degeneration and resulting decline in conception efficiency of bulls.

Vitamin A deficiency can occur in animals fed stored feeds or dried out pastures. Vitamin A deficiency can cause delayed puberty of both sexes, abortion, stillbirth, the birth of weak blind calves and retained placenta. Intake of fresh green feed prevents deficiency.

Lean et al. (2003) have reviewed the impact of nutritional deficiencies on reproductive performance in temperate dairying conditions.

14 Managing abortion

Abortion is one manifestation of the premature death of a conceptus and expulsion of a foetus. Other forms of interrupted pregnancy are early embryonic death that may go unnoticed and be attributed to a failure to conceive, foetal death followed by maceration and absorption, foetal death and mummification and premature birth or stillbirth.

Observed abortions and cases where it is recognised that cows are not pregnant after they have previously been diagnosed pregnant should both be regarded as abortions. In seasonally calving herds, determining the abortion risk as a proportion of confirmed pregnant cows that abort is a satisfactory measure of abortions. The proportion should not exceed 5%, although it can be as low as 0%. Abortion rates are sometimes reported and are variously calculated depending on their intended use. Calculations include the number of abortions occurring in a period as the numerator and the average number of cows pregnant in the period as denominator. The specification of cows in the denominator varies, depending on their pregnancy status and stage, under different definitions of abortion rate.

Abortion is noted when a diagnosed pregnancy is followed by oestrus and then a negative pregnancy diagnosis, when an aborted foetus or foetal membrane is observed or when an abnormal discharge from the vulva is observed after a positive pregnancy diagnosis and on repeat examination, the cow is not pregnant. The observed discharge may be serosanguineous, purulent or contain foetal membranes or their remnants. In the author's experience, 6-7% of pregnancies end in abortion in temperate regions and around 12% in tropical regions.

If abortion is suspected as a herd problem, records should be examined to determine the number and incidence rate of the abortions to confirm that a problem exists. Obtaining a full herd history is required in abortion investigations to aid in diagnosis and to help determine causation, so that remedial and preventive actions can be taken. Possible exposures to infectious causes can be revealed through analysis of animal introductions, events on neighbouring farms, travel of animals, changes in feed and water source. The general health of the herd can be examined and assessed through the herd history as well as the immunisation history and medications used. Corticosteroid and prostaglandin medications are known to cause abortions so their use in the herd should be examined.

Linking epidemiological factors, such as the dates; number of abortions; stage of gestation at abortion; groups affected based on age, location, feeds and water source, can assist in determining causes, as can the identification of the sires of aborted foetuses and an assessment of the fertility of the herd. If feasible, a clinical examination of the whole herd should be undertaken but, in large herds, this may need to be restricted to aborting animals, followed (if possible) by all cows that have been diagnosed as pregnant, in order to identify any cows that have aborted, detect mummies, pyometra and any cows that show signs of illness.

Laboratory examination is highly important in abortion investigations. The diagnostic approaches to infectious causes of reproductive failure, particularly abortion in Australia and New Zealand have been reviewed recently (Reichel et al., 2018). Methods can include diagnostic tests such as PCR and serology for known infectious abortifacient agents, histology and culture of aborted foetal and foetal membrane tissues and examination of feeds and water for infectious agents such as fungi or bacteria. Analyses also can be conducted for toxins. The success of obtaining a diagnosis can be low on submitted samples for several reasons. The necessary cause of the abortion may be gone by the time of abortion and this is even more likely by the time an investigation is undertaken. As an example, feeds containing causal organisms or toxins may have been fully utilised by the time an investigation is undertaken so that there is none of the offending feed left when feed is sampled. Abortion is frequently delayed after foetal death and this can lead to autolysis that masks changes that are characteristic of the cause of abortion. Desired tissues, such as the aborted foetus or foetal membranes, are not always available which limits diagnostic accuracy.

Ideally an entire, fresh foetus can be taken to a diagnostic laboratory for a full examination. If this is not feasible, fresh foetal heart blood, stomach contents and lung samples should be forwarded on ice. Fixed tissue samples of foetal heart, brain, lung, liver and kidney are preferred by pathologists; tissues from cotyledons and caruncles are also preferred. Further useful samples are acute and convalescent maternal serum, samples of feed and water if either is suspected as an associated factor in the abortions, including plant specimens when these are possible causes.

In situations where abortion outbreaks occur, clients are keen to acquire a prognosis for their herd. The prognosis depends on the cause of the abortions, the number of susceptible animals exposed, the stage of pregnancy of animals in the herd and the time between exposure to the abortifacient and diagnosis. In many cases, an outbreak may be almost over before there is veterinary involvement or before a diagnosis is made.

15 Conclusion and future trends

Reproductive health and management programmes have changed over the past 45 years by supplementing the previously used intense monitoring and problem identification strategies with ongoing intervention and control. Hormonal manipulation of the reproductive cycles of cows is becoming far more widespread and is used as the standard method of breeding cows in some situations. This change has been necessitated by ongoing increasing herd sizes, increased milk production per cow and reliance on management by employed labour for vital farm functions such as oestrus detection and selection of cows for artificial insemination. Intensive management programmes are assisted by on-farm computers that are often now linked to cloud-based management systems. Systems now allow the rapid daily review and selection of cows requiring management actions and selection of cows for examination and programmed interventions on veterinary visits.

There is little doubt that as computer artificial intelligence is advanced, machine-based management will increase and cause reduction in the need for inputs by people. Small examples of this exist as in automated oestrus detection with automated diverting of cows to be held for artificial insemination and computer recognition of mastitis and lameness with similar diversion for further examination and treatment.

It is highly likely that treatment regimens delivered by a single device will be developed for oestrus induction and synchronisation. Previous attempts to develop such devices include an intelligent breeding device that delivers hormonal treatments intravaginally to synchronise and induce oestrus through a computer-controlled insert. Micronisation and other developing technologies make such single device treatment programs increasingly feasible and likely. They would offer benefits of precise timing, precise dosing and a decreased chance of environmental contamination or accidental exposure to the veterinary medicines used in hormonal breeding control programs.

16 Where to look for further information

Reproductive health and management of cattle in Australia and New Zealand in seasonal and year-round calving herds are described well in Chapter 11 'Reproduction and disorders of the reproductive system' by Parkinson, T. J., Vermunt, J. J., Malmo, J. and Weston, J. F. in *Diseases of Cattle in Australasia*, Editors Parkinson, T. J., Vermunt, J. J. and Malmo, J. Published by New Zealand Veterinary Association Foundation for Continuing Education (VetLearn), Wellington, New Zealand (2010).

Postpartum anoestrus and its management are extensively reviewed in Chapter 52 'Postpartum anestrus and its management in dairy cattle' by

Ambrose, D. J. in *Bovine Reproduction*, Editor Hopper, R. M. Published by John Wiley and Sons (2015). https://onlinelibrary.wiley.com/doi/book/10.1002/9781118833971.

Reproductive management and changes occurring in herd management of pastured seasonally calving herds are described in 2 papers by Macmillan, K. L. (2002). Advances in bovine theriogenology in New Zealand. 1. Pregnancy, parturition and the postpartum period. New Zealand Veterinary Journal, 50:sup3, 67–73 and 2. Breeding management and technologies for improved reproduction. *New Zealand Veterinary Journal*, 50:sup3, 74–80.

Reproductive management and veterinary intervention have been reviewed using a Hazards Analysis Critical Control Point (HACCP) approach by Lean, I. J., Rabiee, A. R. and Moss, N. (2003). A hazards analysis critical control point approach to improving reproductive performance in lactating dairy cows. Proceedings of the 20th Annual Seminar, Society of Dairy Cattle Veterinarians of the NZVA. Pp. 419-444.

Protocols for synchronisation and re-synchronisation programs for dairy cows are available on the web site of the Dairy cattle Reproduction Council at https://www.dcrcouncil.org/wp-content/uploads/2019/04/Dairy-Cow-Protocol-Sheet-Updated-2018.pdf.

Key journals addressing reproductive management of dairy herds include:

- Animal Reproduction Science. https://www.sciencedirect.com/journal/animal-reproduction-science.
- Journal of Dairy Science. https://www.journalofdairyscience.org/.
- Theriogenology. https://www.journals.elsevier.com/theriogenology.

Congresses addressing dairy cattle reproduction include the two listed below. There are also many national veterinary associations for buiatrics or cattle health and management associated with the World Buiatrics Association that hold national congresses where reproductive health and management are popular topics of presentations:

- International Congress on Animal Reproduction. Normally held every 4 years. http://animalreproduction.org/icar/
- World Association for Buiatrics Congress and its proceedings. Normally held biennially. http://Buiatrics.com

17 References

Ambrose, D. J. (2015) Chapter 52. Postpartum anestrus and its management in dairy cattle. In: *Bovine Reproduction* Hopper, R. M. (Ed.). John Wiley and Sons, Hoboken, NJ.

Anon. (2020). *The InCalf Fertility Focus Users Guide*. DairyNZ, Hamilton, New Zealand. Available at: https://www.dairynz.co.nz/publications/animal/incalf-fertility-focus-users -guide/. Accessed on 28 Sept 2020.

Brito, L. F. C. and Palmer, C. W. (2004). Cystic ovarian disease in cattle. *Large Animal Vet. Rounds* 4(10), 7 pp. Department of Large Animal Clinical Sciences, Western College of Veterinary Medicine, Saskatoon.

Brownlie, T. S., Morton, J. M., Heuer, C., Hunnam, J. and McDougall, S. (2014). Reproductive performance of seasonal-calving, pasture-based dairy herds in four regions of New Zealand. *N. Z. Vet. J.* 62(2),77–86.

Chagas, L. M., Bass, J. J., Blache, D., Burke, C. R., Kay, J. K., Lindsay, D. R., Lucy, M. C., Martin, G. B., Meier, S., Rhodes, F. M., Roche, J. R., Thatcher, W. W. and Webb, R. (2007). Invited review: new perspectives on the roles of nutrition and metabolic priorities in the subfertility of high producing dairy cows. *J. Dairy Sci.* 90(9), 4022–4032.

Denis-Robichaud, J., Cerri, R. L. A., Jones-Bitton, A. and LeBlanc, S. J. (2016). Survey of reproductive management on Canadian dairy farms. *J. Dairy Sci.* 99(11), 9339-9351.

Dubuc, J., Duffield, T. F., Leslie, K. E., Walton, J. S. and LeBlanc, S. J. (2010). Risk factors for postpartum uterine diseases in dairy cows. *J. Dairy Sci.* 93(12), 5764-5771.

Fetrow, J., McClary, D., Harman, R., Butcher, K., Weaver, L., Studer, E., Erlich, J., Etrherington, W., Guterbock, W., Klingborg, D., Reneau, J. and Williamson, N. (1990). Calculating selected reproductive indices: recommendations of the American Association of Bovine Practitioners. *J. Dairy Sci.* 73(1), 78-90.

Fielden, E. D., Macmillan, K. L. and Watson, J. D. (1973). The anoestrous syndrome in New Zealand dairy cattle. 1. A preliminary investigation. *N. Z. Vet. J.* 21(5), 77-81.

Fricke, P. (2019). Evolution of fertility programs for lactating cows. *Clin. Theriogenol.* 11, 317-328.

Gyawu, P., Ducker, M. J., Pope, G. S., Saunders, R. W. and Wilson, G. D. (1991). The value of progesterone, oestradiol benzoate and cloprostenol in controlling the timing of oestrus and ovulation in dairy cows and allowing fixed time insemination. *Br. Vet. J.* 147(2), 171-182.

Lawton, D. E. B., Mead, F. M. and Baldwin, R. R. (2000). Farmer record of pregnancy status pre-slaughter compared with actual pregnancy status post-slaughter and prevalence of gross genital tract abnormalities in New Zealand dairy cows. *N. Z. Vet. J.* 48(6), 160-165.

Macmillan, K. L. (2002). Advances in bovine theriogenology in New Zealand. 1. Pregnancy, parturition and the postpartum period. *N. Z. Vet. J.* 50(3) (Suppl.),67–73.

McDougall, S. (1994). Postpartum anoestrum in the pasture grazed New Zealand dairy cow. Ph.D. Thesis, Massey University, Palmerston North.

McDougall, S. and Compton, C. (2008). 'Effects of treatment of 'not detected in oestrus' cows with gonadotrophin releasing hormone, prostaglandin and progesterone. Proceedings of the Society of Dairy Cattle Veterinarians of the NZVA, Pp 35-50.

McDougall, S., Leijnse, P., Day, A. M., Macmillan, K. L. and Williamson, N. B. (1993). A case control study of anoestrum in New Zealand dairy cows. Proceedings of the New Zealand of Society of Animal Production 53, 101-103.

Mwaanga, E. S. and Janowski, T. (2000). Anoestrus in dairy cows: causes, prevalence and clinical forms. *Reprod. Domest. Anim.* 35(5),193-200.

Pursley, J. R., Mee, M. O. and Wiltbank, M. C. (1995). Synchronisation of ovulation in dairy cows using PGF2α and GnRH. *Theriogenology* 44(7), 915-923.

Rabiee, A. R., Lean, I. J. and Stevenson, M. A. (2005). Efficacy of Ovsynch program on reproductive performance in dairy cattle: a meta-analysis. *J. Dairy Sci.* 88(8), 2754–2770.

Reichel, M. P., Wahl, L. C. and Hill, F. I. (2018). Review of diagnostic procedures and approaches to infectious causes of reproductive failures of cattle in Australia and New Zealand. *Front. Vet. Sci.* 5(222), 15 pp. Available at: https://www.ncbi.nlm.nih.gov/pmc/articles/PMC6176146/. Accessed on 28 Sept 2020.

Stevenson, J. S. and Britt, J. H. (2017). A 100 year review: practical female reproductive management. *J. Dairy Sci.* 100(12), 10292–10313.

Stevenson, J. S., Hill, S. L., Nebel, R. L. and DeJarnette, J. M. (2014). Ovulation timing and conception risk after automated activity monitoring in lactating dairy cows. *J. Dairy Sci.* 97(7), 4296–4308.

Williamson, N. B. (1987). The interpretation of herd records and clinical findings for identifying and solving problems of infertility. *Compend. Contin. Educ. Pract. Vet.* 9(1), F15, 14–23.

Williamson, N. B. (2005). Heat detection. Proceedings of the Society of Dairy Cattle Veterinarians of the NZVA, Pp 141–151.

Williamson, N. B., Morris, R. S., Blood, D. C. and Cannon, C. M. (1972). A study of oestrous behaviour and oestrus detection methods in a large commercial dairy herd. *Vet. Rec.* 91(3), 50–58.

Zemjanis, R., Fahning, M. L. and Schultz, R. H. (1969). Anestrus. The practitioners dilemma. *Vet. Scope* XIV, 14–21.

Chapter 2

Advances in understanding behavioral needs and improving the welfare of calves and heifers

Emily Miller-Cushon, University of Florida, USA; and Jennifer Van Os, University of Wisconsin-Madison, USA

1 Introduction

Management of dairy calves and heifers is critical, having the potential to dramatically influence animal welfare and performance during a sensitive period of life and exert longer-term effects on outcomes into lactation. Despite the acknowledged importance of calf rearing, management of dairy calves and heifers is highly variable across the dairy industry. Much research in recent years has furthered our knowledge of calf behavior, having important implications for approaches to manage youngstock housing to improve welfare.

Historically, calf and heifer housing and management decisions have been driven by economic considerations and the welfare and sustainability issue of calf morbidity and mortality in early rearing. However, there have been recent shifts in perspective on calf rearing, including increasing awareness of the influence that early life experiences exert on development, having longer-term implications for both animal productivity and welfare. As such, calf management decisions in recent years have begun to make changes with both short-term and longer-term outcomes in mind. For example, calf milk allowance has historically been restricted to encourage early weaning onto solid feed; yet the practice of

http://dx.doi.org/10.19103/AS.2020.0084.10

providing more biologically appropriate levels of milk during the first weeks of life is becoming more common, in light of evidence of long-term production and health benefits associated with increasing early milk allotment (as reviewed by Khan et al., 2011a).

In addition to increasing longer-term perspectives on optimal housing, calf management decisions are increasingly made with consideration of behavioral needs in mind, due to concern for aspects of animal subjective experience in addition to basic health. This mirrors a shift across all of animal agriculture, where increasing public and industry concern for animal welfare, and research focused on understanding animal behavior and welfare, has been given more priority.

In dairy calves and heifers, consideration of behavioral needs has encompassed aspects of social interactions and feeding, rest and comfort, and opportunities for other behavioral expression, such as grooming and play. More generally, there is growing understanding of the importance of environmental complexity and opportunities for behavioral expression and the role of these early experiences in improving lifelong welfare and reducing behavioral problems, such as the development of abnormal behaviors. Further, understanding of animal behavior and refinement of management to meet behavioral needs increasingly occurs at an individual level, and research in this area reveals individual variability in behavioral development and preferences that should be accommodated to improve welfare. Overall, developing research in this area provides a means to refine management for dairy calf and heifers to improve short- and long-term welfare.

2 Addressing social needs

Under natural conditions, the dam will seek isolation from the herd before calving (Lidfors et al., 1994) such that the calf's initial social contact includes suckling and maternal grooming (Jensen, 2011). Following the return of the calf and cow to the herd, observations of calves reared on pasture suggested that they rest and interact with other calves within the first weeks of life (Sato et al., 1987) and, in natural conditions, long-lasting bonds develop between a female and her female offspring, as well as other females (Reinhardt and Reinhardt, 1981). In the dairy industry, it is common practice to separate the calf from the cow shortly after birth, and subsequent calf housing practices vary widely in social environments and exert broad and long-term effects on behavioral development.

2.1 Maternal contact

Cow-calf separation shortly after birth is a central practice in the dairy industry; yet it is increasingly under scrutiny due to societal concerns that separation negatively affects animal welfare (Ventura et al., 2013). Research in this area has evaluated welfare consequences associated with the timeframe of separating the calf and cow, and recent research efforts have explored approaches and implications to facilitating prolonged cow-calf contact.

It is well established that calves and cows exhibit acute behavioral responses to separation, including vocalizations and increased movement, and evidence suggests that earlier separation (within 24 h, compared to 4 or more days) may reduce vocalizations and time spent looking out of the pen (as reviewed by Meagher et al., 2019). This suggests that distress upon separation may relate to strength of the established bond, and has been used as a basis for recommending prompt cow-calf separation. However, recent research has also evaluated possible behavioral and performance benefits of prolonged maternal contact (Meagher et al., 2019).

There is evidence of bonding in cow-calf pairs, even when suckling is prevented (Johnsen et al., 2015), suggesting that longer-term contact may have a range of developmental benefits for the calf. Prolonged maternal contact supports social development, such as promoting the appropriate expression of submissive behavior (Buchli et al., 2017) and increased expression of play behavior (Flower and Weary, 2001). However, some of these social benefits are not unique to maternal contact and are also seen in calves housed with peers (as discussed in the following section). Prolonged maternal contact is also generally found to reduce development of abnormal oral behaviors, such as cross-sucking and tongue-rolling, compared to calves housed socially but separated from the dam (Roth et al., 2009; Fröberg et al., 2007), which may indicate specific benefits of suckling.

In light of behavioral benefits and societal concerns regarding cow-calf separation, various systems to facilitate prolonged cow-calf contact are being developed. These include free contact, restricted suckling or partial day contact, and use of a nurse or foster cow (Johnsen et al., 2016). However, prolonged cow-calf contact is not currently considered widely feasible, and it remains most common for dairy calves to be housed apart from the dam from shortly after birth.

2.2 Social housing for calves during the pre-weaning period

Once separated from the dam, typical housing systems for dairy calves during the pre-weaning period vary across farms. Calves may be housed individually, which remains most common in many regions, including Canada (Vasseur et al., 2010), the United States (USDA, 2016), and Europe (Marcé et al., 2010), or with same-age social contact, which can take various forms from housing in pairs to larger group pens of calves. Individual rearing became the standard practice in part due to concerns about calf morbidity and mortality, with social isolation seen as reducing the risk of direct calf-to-calf disease transmission and allowing for ease of individual monitoring before the advent of modern automated technologies. However, poor health and performance sometimes observed in large groups of calves was often confounded by low milk allowance (resulting in insufficient nutrient intake and reduced immune function; reviewed by Kahn et al., 2011a) and poor ventilation in older facilities, along with other management risk factors such as poor sanitation, bedding, and dynamic (as opposed to all-in/all-out) grouping practices. Evidence suggests no clear relationship between social housing and disease, whereas management factors such as hygiene, ventilation, nutrition, and calf monitoring are critical with respect to health (as reviewed by Costa et al., 2016a). Rather, there is now a well-established body of evidence indicating broad benefits of social contact during the critical pre-weaning period.

In the last decade, numerous studies have shown that rearing pre-weaned calves in pairs or small groups has many benefits for both animal welfare and performance (Costa et al., 2016a). Social housing improves social and cognitive development, with calves showing improved behavioral flexibility (Gaillard et al., 2014) and willingness to try new feeds (Whalin et al., 2018) when reared in social housing, compared with those reared individually. This translates into better resilience to weaning stress and improved performance during this period (Chua et al., 2002; de Paula Vieira et al., 2010; Duve et al., 2012), although evidence of longer-term effects on behavior and welfare are somewhat less studied to date. Across studies (particularly when fed higher milk allowances), socially housed calves perform at least as well as individually housed calves in terms of solid feed intake, bodyweight at weaning, and average daily gain. Additionally, housing calves in groups while maintaining at least the same space per animal results in an increase in total space, allowing for the expression of a greater range of natural behaviors, including play (Jensen et al., 2015).

Social contact has clear implications for calf affective state. There is broad evidence that calves prefer social contact and seek it out. For example, a study characterizing behavior of pre-weaned calves housed in adjacent outdoor hutches, with a shared outdoor enclosure, found that calves spent over 80% of the indoor time inside the same hutch (Wormsbecher et al., 2017). Although this study was not designed as a formal preference test, the calves' choice to spend time inside the same 2.4 m^2 hutch, at the expense of a larger individual space allowance, suggests that social proximity was a high priority for them. Calves are motivated for social contact, based on willingness to perform an operant task to gain access to another calf, even when only head contact through metal bars was allowed (Holm et al., 2002), suggesting that providing social contact is accommodating a behavioral need. In recent findings, pair-housed calves were found to respond more positively to ambiguous cues, presented in a spatial go/no-go task, compared to individually housed calves (Bučková et al., 2019), demonstrating a more optimistic judgment bias indicative of improved welfare.

Many studies in this area have compared individual and pair housing, suggesting that benefits of social contact are achieved when calves access only a single social partner. However, the needs of calves within larger and more complex social groups have been less explored. A better understanding of social interactions within larger groups of calves may be a means to evaluate individual welfare or refine housing to better accommodate individual behavioral needs. For example, the amount and complexity of physical space provided to group-housed calves may be important for the development and maintenance of social relationships, such as allowing for avoidance of dominant animals or social withdrawal. Social proximity is widely considered an indicator of social bonding, and varies positively with familiarity of social companions (Færevik et al., 2006). Preference for social proximity or isolation, however, may vary between individuals and depend on various factors, including individual personality (Lecorps et al., 2018).

Changes in social proximity have been observed to coincide with morbidity in dairy calves (as modeled using a respiratory disease challenge; Hixson et al., 2018), although it is not well-understood whether changes in social proximity may be due to motivation for social withdrawal or avoidance of a sick calf by pen-mates. Preference for social contact may be better accommodated in group-housed calves through increasing environmental complexity. For example, physical barriers allow for cows to choose to self-isolate during calving (Proudfoot et al., 2014). Recent evidence suggests that dairy calves may seek isolation while in pain, as expressed by increased use of a physical barrier providing visual separation from pen-mates following disbudding (Gingerich et al., 2020). Variability in social behavior between individuals and over time, and welfare implications of accommodating broader expression of social behaviors in the home pen, remain an area of developing research.

2.3 Social dynamics and transitions in weaned heifers

Following weaning, it is common practice on many farms to move heifers to group-housing. For calves previously reared individually, this movement to a social group is a stressful transition that coincides with weaning, such that the separation of these events is preferable (Weary et al., 2008). Housing management during the pre-weaning period may ease this transition, as there is abundant evidence that early life experience has a longer-term effect on behavioral development of dairy calves. Calves reared socially react less fearfully in novel environments (Jensen et al., 1997), indicating an improved ability to adapt to the post-weaning environment. Dairy heifers experience a range of social and housing transitions as they develop, including social regroupings, movement to a freestall barn (as discussed in Section 5.2 of this chapter), and introduction to the milk parlor. Consideration of their behavioral needs through these periods of adaptation and learning may have important welfare implications.

3 Addressing feeding needs

The dairy industry is unique among intensively managed livestock in separating calves at birth from the dam as standard practice. As such, nutritional management from birth is human-controlled, and there are a wide range of approaches for raising calves during their period of consuming milk (or milk substitute) and for providing solid feed and weaning calves from milk.

3.1 Milk feeding of calves

Research concerning needs of dairy calves surrounding milk-feeding has focused on volume, feeding method, and transition from milk to solid feed. The dairy industry has a history of providing dairy calves with restricted milk allotments (less than half of the volume the calf would consume if suckling the cow, or provided free access to milk or milk replacer; Appleby et al., 2001; Jasper and Weary, 2002), with the aim of reducing costs associated with milk-feeding and encouraging solid feed intake to facilitate early weaning. Restricted milk allowances are typically provided at a limited frequency (2-3 times/d), in contrast to feeding which is distributed across the day (Fig. 1) at frequencies of 7-10 times/day for calves with free access to milk (Miller-Cushon et al., 2013a), which mirrors suckling patterns of calves with the cow (reviewed by de Passillé, 2001). Increasing adoption of higher milk allowance feeding plans has been facilitated by the growing use of automated milk feeders, which also provide milk at flexible times, resulting in meal patterning that, in theory, more closely resembles that of a suckling calf. A little-explored potential downside of automatic milk-feeding systems, however, is the restriction of synchronized feeding when only one feeding station is typically provided to groups of 20 or more calves.

   A growing body of evidence provides support for more biologically appropriate milk feeding volumes, due to improved performance (including growth, solid feed intake, rumen development, and health, as reviewed by Khan et al., 2011a) and alleviation of the negative affective state of hunger, as indicated by a variety of behavioral responses. Calves vocalize less frequently when provided more milk (Thomas et al., 2001), spend more time lying down and displace other calves from the feeder less (de Paula Vieira et al., 2008), and have increased locomotor play (Jensen et al., 2015). The provision of high milk allowances is appropriate from birth, a time period when the calf has not begun consuming energy from solid feed yet may ingest large quantities of milk (e.g.

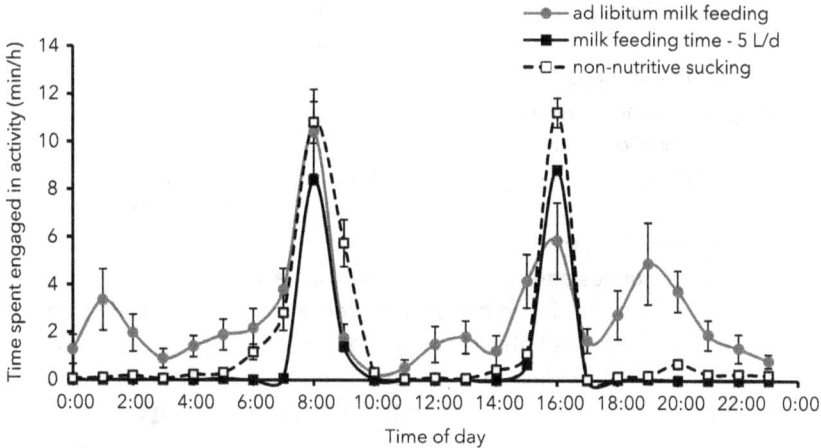

Figure 1 Hourly feeding time for calves provided milk, free access to milk replacer, and calves with restricted access (5 L/d, provided in two meals at 0800 h and 1600 h) and time engaged in teat-directed non-nutritive sucking for calves provided restricted amounts of milk. (Adapted from Miller-Cushon et al., 2013a; reproduced with permission from the Journal of Dairy Science).

averaging 8.6 kg/d in the first week of life, nearly double the allocation of calves provided milk at a conventional restricted rate of 10% of body weight; Appleby et al., 2001).

In addition to milk allowance, feeding methods have consequences for calf behavior and welfare. Obtaining milk through a teat is considered to be rewarding to calves, and has therefore been commonly used as a positive stimulus in studies where calves are placed in testing scenarios (e.g. Horvath et al., 2017; Ede et al., 2018). In practice on some farms, milk allowances are provided in buckets (or trough), requiring the calf to drink or sip the milk, rather than via an artificial teat which accommodates suckling behavior. In most cases, but not all, bucket feeding goes hand in hand with providing restricted milk allowance. It is well established that having an outlet for natural suckling behavior is an important need for calves, with the action of suckling having the function of stimulating hormones involved in digestion and satiety (de Passillé et al., 1993), whereas calves consuming milk from buckets have higher heart rates and increased non-nutritive oral behaviors (Veissier et al., 2002).

In general, performance of non-nutritive oral behaviors may depend on a range of management factors, as discussed throughout this chapter, but are closely related to milk feeding methods and allowance. Cross sucking in particular is considered a nuisance behavior in group-housed calves, with the potential to result in frostbitten ears (in the case of outdoor-housed calves in cold weather) or damaged teats, mastitis, and subsequently reduced milk production (Lidfors and Isberg, 2003); yet not all research indicates that cross-sucking affects udder health (Vaughan et al., 2016). Feeding milk through a teat instead of open buckets can reduce the incidence of cross sucking (de Passillé, 2001; Loberg and Lidfors, 2001; Jensen and Budde, 2006), as can providing a non-nutritive 'dummy' teat to redirect suckling behavior away from conspecifics (de Passillé, 2001; Lidfors and Isberg, 2003), including when calves are fed milk from an open bucket (Veissier et al., 2002). Non-nutritive suckling, including cross sucking, has a periodicity such that the majority of suckling behavior is stimulated when calves taste milk or milk replacer, and this behavior continues after milk is ingested (de Passillé, 2001). Non-nutritive sucking throughout the day in calves provided a restricted milk allowance is shown in Fig. 1, where this behavior occurs for the longest periods around milk meals and surpasses feeding time. To effectively reduce cross sucking, milk-feeding equipment with teats or non-nutritive teats need to remain accessible to the calf for at least 20 min after milk ingestion (Jung and Lidfors, 2001).

In group-housed calves, general social dynamics surrounding milk feeding are important to accommodate. Calves exhibit a high degree of feeding synchrony, even choosing to feed adjacent to a pen-mate where feeding space is limited, despite availability of feeding space elsewhere (Miller-Cushon and DeVries, 2016). This suggests that feeders that thwart social feeding, such as automated milk feeders which provide a stall for one calf, may frustrate feeding motivation. A reduction in feeding synchrony and a high degree of competition is evident even in pair-housed calves provided milk *ad libitum* via a single teat (Miller-Cushon et al., 2014a), yet autofeeders with a single feeding space often accommodate approximately 20 calves on-farm (Jorgensen et al., 2017). Social elicitation of feeding behavior by suckling pen-mates may also explain why cross-sucking has been observed by farmers who provide even high milk allowances to group-housed calves fed by an automated milk feeder, and more research could help elucidate this possibility. Problematic social interactions surrounding feeding, such as displacements, can also be reduced through increasing pen complexity, such as the provision of simple barriers to reduce competition for access to milk (Jensen et al., 2008). However, social feeding appears to be preferred in young calves, and so approaches to facilitate it may be beneficial.

3.2 Weaning from milk

Dairy calves are weaned from milk around 6-8 weeks of age, many months earlier than may occur under more naturalistic settings (e.g. 6 months for beef calves), and the timeframe and method of weaning have important behavioral and performance implications for the dairy calf.

There has been attention placed on refining feeding plans to strike a balance between milk allowance and weaning requirements. Calves provided more milk consume less solid feed prior to weaning, delaying rumen development, and may not maintain consistent growth if weaned too abruptly (reviewed by Khan et al., 2011a). Improvements in growth are seen with gradual reductions of milk allowance (Khan et al., 2007; Sweeney et al., 2010), as well as initiating weaning later (e.g. Eckert et al., 2015; Mirzaei et al., 2018). Cross sucking has been observed to increase directly after weaning, presumably in response to a drop in energy intake (de Passillé et al., 2010), and calves who are better established on solid feed are less likely to cross suck at this time (Keil and Langhans, 2001). Timeframe of weaning impacts both performance and behavior, with calves weaned earlier from a higher milk allowance having more frequent vocalizations and non-nutritive oral behavior (e.g. 6 weeks of age; Eckert et al., 2015).

In recent years, there has been interest in understanding and accommodating the individual needs and behavioral preferences of calves surrounding feeding in the first months of life. Whereas animals are usually managed at a group level, with diets formulated for average nutritional requirements and protocols adopted based on average needs, individuals vary in feeding behavior and preferences. Milk intake and voluntary consumption of solid feed is one area where this variability is particularly apparent. Calves exhibit a wide range of milk intakes and meal patterns when provided free access to milk. Appleby et al. (2001) described *ad libitum* milk intake ranging from 6 to 17 Ld during week 4 of life, with meal frequency ranging from 3 to 22 meals/d. Correspondingly, calves provided milk *ad libitum* also vary in their initiation of solid feed consumption (ranging from 0.01 to 0.31 kg/d during week 4 of life; Appleby et al., 2001). Individual variability in voluntary starter intake during the pre-weaning period has important implications for performance through weaning. Monitoring of individual feeding behavior traits is increasingly facilitated by technology, such as automated milk and starter feeders, and provides opportunity to manage calves at an individual level and accommodate individual differences.

Weaning based on individual starter intake, rather than age, provides a promising approach to individualize management, with potential to reduce cross-sucking during weaning and improve weight gain (Roth et al., 2008). These methods may be particularly useful when calves are provided higher milk allowances. In one study, weaning by reducing a 12 L/d milk allotment as the calf progressively met targets for starter intake (200, 600, 1000, and 1400 g/d; de Passillé and Rushen, 2016) yielded a wide range in duration of weaning (7-49 d) and ages at weaning (58-94 d). Compared to calves weaned by age, this intake-based weaning resulted in reduced milk intake with maintained growth (de Passillé and Rushen, 2016). Benetton et al. (2019) also found that a similar intake-based weaning protocol reduced total milk consumption without negatively impacting growth, compared with weaning based on age. However, those authors also reported that nearly 40% of calves failed to meet some or all of the starter intake targets by the age limit imposed and calves that weaned by intake later (after 63 d) had reduced body weight post-weaning. These findings confirm wide individual variability in early development of feeding behavior and suggest that some calves may still struggle during a weaning transition at this age range, even when their differences in voluntary solid feed intake are somewhat accommodated.

Recent research has further explored predictors of individual variability in behavior outcomes, and evidence suggests that individual variability in feeding behavior through weaning may be explained by differences in personality. Neave et al. (2018) found that increased exploration and activity, as assessed through novel environment, human approach, and novel object tests, was associated with earlier acceptance of solid feed and increased intake through weaning. Evidence of considerable variability in feeding behavior, and associations between feeding behavior and personality traits, indicate that that refining management to meet animal needs necessitates an understanding of individual behavior.

Social environment during the pre-weaning period has implications for acceptance of solid feed and performance through weaning. Social contact during the pre-weaning period stimulates feed intake and reduces feed neophobia (reviewed by Costa et al., 2016a), indicating a function of social contact regarding the development of feeding behavior. Social contact is likely of particular importance for acceptance of solid feed in calves provided higher milk allowances where weaning remains a challenge. In one study, pair-housing stimulated solid feed intake and supported increased growth through weaning in calves provided more milk, whereas pair-housing did not affect performance in calves provided less milk (9 vs. 5 L/d; Jensen et al., 2015). Some evidence suggests that this effect may be enhanced through the provision of more experienced social models. For example, De Paula Vieira et al. (2012) found that housing calves in groups that included an older weaned calf spent more time at the feed bunk and gained more weight prior to weaning. Whereas animals are typically housed by age, especially for pre-weaned calves where vulnerability to disease is a common concern, providing contact with more experienced social models may provide a mechanism to ease dietary and housing transitions that warrant further exploration in weaned heifers.

3.3 Solid feed intake and preferences in young calves

In natural conditions, calves begin to investigate the pasture and consume small amounts in the presence of the dam during the first weeks of life (Nicol and Sharafeldin, 1975). Ruminants have developed the ability to select and learn about a wide range of feed types (Provenza and Balph, 1987), yet under conventional intensive management dairy calves and heifers are typically provided a single mixed diet formulated from a range of ingredients, either pelleted or mixed, potentially alongside some form of forage. Behavioral considerations for calves regarding solid feed provision include palatability and sensory factors, acceptance or selection of feed, variety of feed types provided, and social factors surrounding feeding.

Dairy calves exhibit clear preferences for certain feed ingredients and flavors, which have been studied primarily with the aim of formulating rations that will encourage greater feed intake. For example, in a series of experiments, calves exhibited a consistent preference for soybean meal over other protein sources, increasing their intake of this ingredient in pairwise preference tests and sorting within a mixed diet in favor of that component (Miller-Cushon et al., 2014b). There is no clear evidence to suggest that food palatability influences intake over longer time frames, but there has been some interest in using flavor additives to encourage acceptance of solid feed. Montoro et al. (2011) found that flavoring a starter pellet to match milk replacer during the week prior to weaning had a slight effect in stimulating intake in calves with the lowest pre-weaning intake, suggesting that sensory properties of feed might be influential for calves with lower intakes or, potentially, increased neophobia. In general, the role of sensory properties of feeds in encouraging transition to novel diets has not been well explored.

The variety of feed types and flavors provided to young calves may also be important to consider, to accommodate both individual variability in perceived palatability and possible preference for variety. Cows have more taste receptors on their tongues than humans (Roura et al., 2008), indicating that they are highly sensitive to flavor. In modern dairy production, there is a focus on optimal nutritional formulation of relatively uniform diets, which provide for average nutrient and energy needs, but do not accommodate individual preference or provide opportunity for variety, choice, or sampling and exploratory behaviors. Despite clear evidence of preference for particular flavors and feed ingredients, calves maintain intake of less preferred feed types when offered a choice in experimental settings (Miller-Cushon et al., 2014b). Meagher et al. (2017) also found that weaned heifers selected a fraction of their diet from a rotating offering of a second forage variety or flavored ration, with preference for the varied option subject to individual variability.

This sampling behavior has been attributed to an adaptive function of monitoring the environment (Kyriazakis and Oldham, 1993), but there is also evidence to suggest that ruminants prefer a varied diet. Scott and Provenza (1998) found that lambs preferred a variety of flavors when nutrient content between foods was identical, suggesting that food sensory diversity may have hedonic value. Further, lambs able to choose between feeds that varied over time had lower cortisol levels after feeding than lambs provided a single balanced diet (Catanese et al., 2013). Individual variability in flavor preferences is also well-documented in different species (reviewed by Favreau-Peigné et al., 2013), suggesting that provision of a variety of feed types may better accommodate individual preferences and reduce variability in feed intake. Overall, providing a variety of feed types or flavors may accommodate natural foraging behavior, with potential to improve affective state and stimulate feeding in young calves.

Beyond sensory characteristics of feed, nutrient composition, particle length, and the variety of feed types provided may be important from a behavioral and developmental standpoint. Evidence in young ruminants suggests an ability to associate sensory properties of food with nutrient content, such that provision of a choice of foods (e.g. 'cafeteria' style) may allow individuals to select a balanced diet to meet their needs. For example, lambs appear to select from feed choices to avoid excess protein intake (Kyriazakis and Oldham, 1993). To date, there is limited evidence to suggest that young dairy calves are adept at selecting a balanced solid feed diet from a variety of energy sources to meet specific nutritional demands. For example, calves offered a range of six ingredients were found to over-consume protein (Montoro and Bach, 2012). Similarly, calves also had different protein intakes depending on the protein source, with diet selection apparently driven by differences in perceived palatability rather than nutrient composition (Miller-Cushon et al., 2014c). Whereas dietary selection of solid feeds may not be sensitive to nutrient imbalances in young milk-fed calves, selection of milk relative to high-energy concentrate and forage may be driven by individual preferences and needs. Webb et al. (2014a) described wide individual differences in voluntary intake of bull calves offered free access to milk replacer, concentrate, and various forages, and concluded that diets formulated for average needs may differ from preferences exhibited by most calves. Milk-fed dairy calves do exhibit consistent selection for forage, when offered alongside (shown in Fig. 2) or mixed into starter, and selection of forage may depend on dietary requirements.

Dairy calves are adept at sorting within mixed diets even when particles are small (e.g. diets consisting of different sized pellets, Miller-Cushon et al., 2014c) and will sort mixed diets of forage and starter from the first weeks of life (Miller-Cushon et al., 2013b). Research describing sorting behavior of young calves suggests a motivation to consume forage during the pre-weaning period. For example, calves sorted a mixed diet of 70% texturized grain starter and 30% chopped hay in favor of hay during the pre-weaning period (Miller-Cushon et al., 2013b). Although the individual degree of hay selection has not been associated with rumen environment in young calves, it is well established that provision of hay increases rumen pH and stimulates rumen development, compared to calves provided starter only (Khan et al., 2011b), suggesting that sorting in favor of forage may be motivated by post-ingestive feedback

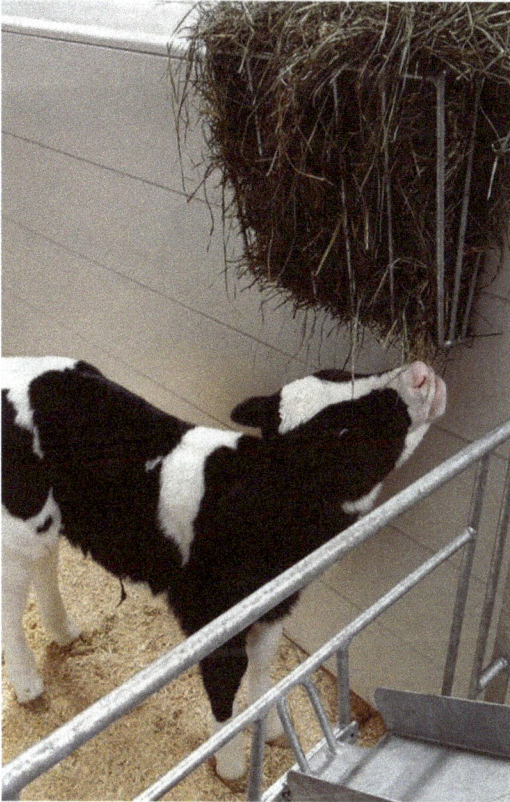

Figure 2 Image of a pre-weaned calf provided with unchopped hay on a commercial dairy farm in British Columbia, Canada. Photo by Jennifer Van Os.

associated with improved gut health. In adult cattle, it is well established that forage selection (Keunen et al., 2002) and sorting in favor of long ration particles (DeVries et al., 2008, 2014) both increase in response to low rumen pH, suggesting a functional outcome associated with this dietary selectivity. Forage intake in young calves may also be motivated by the sensory process of chewing and ingesting longer feed particles. Dairy calves provided hay have reduced abnormal oral behaviors, such as pen-directed sucking (Fig. 3; Haley et al., 1998; Horvath and Miller-Cushon, 2017), suggesting that consuming hay satisfies a behavioral need for oral manipulation and foraging behavior. Similarly, veal calves provided a diet containing grain and hay performed less tongue rolling compared to calves provided milk only, adjusted to maintain similar weight gain (Webb et al., 2012).

Selection in favor of starter relative to hay increases after weaning off of milk (Miller-Cushon et al., 2013b), suggesting that dairy calves may adjust

Figure 3 Image of a weaned heifer engaging in pen-directed sucking at the University of Florida heifer-raising facility in Alachua, Florida, United States. Photo by Katie Gingerich, University of Florida.

dietary selection to accommodate increasing energy demands when milk is no longer provided. Similarly, weaned bull calves sorted a forage-based total mixed ration in favor of long forage particles and against small ration particles when provided a separate supplemental concentrate, and switched their pattern of sorting in favor of smaller higher energy particles when the supplemental concentrate was no longer provided (Costa et al., 2016b). These results suggest that calves adjust sorting behavior, in favor of particles higher in fiber or energy, depending on their nutrient requirements.

3.4 Feeding behavior in weaned heifers

As they grow, dairy heifers are challenged with many dietary and management changes. Feeding behavior patterns developed early in life have potential to influence longer-term feeding behavior and feed preferences (reviewed by Miller-Cushon and DeVries, 2015), suggesting that feed characteristics during critical pre-weaning and weaning periods may have implications for heifer behavior and welfare, with potential to reduce neophobia and influence meal patterns. For example, in recently weaned heifers, the extent of sorting within a mixed diet in favor of preferred feed components is influenced by the opportunity to perform sorting behavior during the pre-weaning period (Miller-Cushon et al., 2013b). In general, the limited work in this area suggests a range of possible carryover effects of the pre-weaning period on post-weaning feeding behavior, which may affect individual responses to dietary and feed management factors.

As discussed in Section 3.2, social housing during the pre-weaning period also affects development of feeding behavior, enhancing acceptance of novel feeds (Costa et al., 2014), reducing latency to begin feeding once grouped after weaning (De Paula Vieira et al., 2010), and improving competitive success for access to feed (Duve et al., 2012); this therefore has clear implications for the welfare of weaned heifers during key dietary and management transitions. Competition surrounding feeding in the post-weaning environment may also relate to previous experience, as evidence suggest that calves subjected to competitive pressure for access to milk continue to engage in more frequent competitive displacements post-weaning (Miller-Cushon et al., 2014a).

Implications of feeding management for behavior and welfare of weaned heifers have generally received less research attention than calves during the milk-feeding stage. However, an important area of consideration is feeding level and forage provision, as it relates to feeding time and the development of abnormal oral behaviors. For adult cattle, the natural feeding-behavior repertoire includes appetitive components relating to foraging (i.e., searching, selecting) and consummatory aspects to ingesting feed (i.e., chewing, ruminating). In natural settings, cattle graze for a period of time ranging between 7 and 13 h/d (reviewed by Kilgour, 2012). However, intensive housing for cattle typically provides a higher energy, readily accessible diet; consequently, feeding time for confined, *ad libitum*-fed lactating cattle is reported to be between 3 and 5 h (Dado and Allen, 1994; Hosseinkhani et al., 2008).

Although not a standard practice, some work in recent years has evaluated the practice of providing limited amounts of high-energy diets (or 'limit-feeding'), which supports normal growth but decreases feeding time along with manure volume. Limit-fed heifers spent less than 90 min/d feeding in one study (Kitts et al., 2011), and only 28% of the time they would normally spend grazing in another study (Redbo, 1990). High-energy, short-particle diets require less chewing, resulting in less saliva production, which normally helps buffer rumen pH (Beauchemin et al., 2008). Consequently, highly fermentable grain-based diets can result in an excessive level of acid in the rumen, which is not effectively ameliorated by saliva secretion (reviewed by Owens et al., 1998).

Furthermore, feed restriction can affect not only physiological aspects of welfare, but also behavioral outcomes. Tongue rolling in heifers is closely related to feed restriction. For example, Redbo and Nordblad (1997) found that the prevalence of tongue rolling in a group of heifers fed *ad libitum* increased from 33% to 92% when they were fed restricted quantities of a high-energy, low-forage diet. The heifers who exhibited tongue-rolling prior to the period of feed restriction also increased the frequency of tongue-rolling bouts when roughage was restricted. Lindström and Redbo (2000) also determined that, regardless of rumen fill, the ability to forage decreased tongue-rolling, indicating the importance of active foraging behavior for developing heifers. This mirrors effects seen in calves, discussed above, where the provision of forage reduced pen-directed sucking. Behaviour related to feed searching is more common before than after oral stereotypies and tongue-rolling occurs primarily in the hours immediately following a meal (Redbo, 1990), further suggesting a relationship between feeding motivation and tongue-rolling.

The ability to perform the appetitive and consummatory components of feeding behavior may be important for post-weaned cattle. When adult dairy cows' rumens were filled through a fistula (and thus their nutritional needs were met and the sensation of gut fill was achieved, but feeding behavior was limited), they still investigated the empty feed bunk with their noses and tongues as they would during the appetitive component of feeding behavior (Lindström and Redbo, 2000). Cattle may be intrinsically motivated to perform feeding-related behaviors, which are more naturally addressed with high-forage diets or grazing. When fed high-energy, grain-based diets, 7-week-old veal calves, 9-month-old dairy heifers, and 13-month-old beef heifers were motivated to obtain forage, and they performed work by pressing panels and pushing weights, respectively (Webb et al., 2014b; Greter et al., 2015; Van Os et al., 2018). Furthermore, heifers with *ad libitum* access to forage also pushed weights to access additional amounts of the same forage, demonstrating the concept of contrafreeloading (Van Os et al., 2018); this may, in part, reflect an inherent motivation to express foraging behavior, independent from satiation or nutrient requirements.

4 Addressing resting and environmental comfort needs

Lying down is well recognized as an important behavioral need for cattle of all ages. Aspects of the housing environment affecting thermal comfort and the physical comfort of the resting area have important implications for cattle welfare at all life stages.

4.1 Rest and comfort in young calves

Calves spend the majority of their day resting (e.g. 17-18 h/d lying down; Hänninen et al., 2005; Camiloti et al., 2012; Bonk et al., 2013). The type and amount of bedding can influence resting behavior. Calves show clear preferences for dry sawdust over wet sawdust, and an aversion to lying on concrete (Camiloti et al., 2012) and prefer lying on sawdust compared to sand or stones (Worth et al., 2015). Duration of time spent lying has been associated with weight gain (Hänninen et al., 2005), and may depend on bedding comfort. Calves provided wood shavings compared to quarry stones spent more time lying (Sutherland et al., 2013), whereas bedding type in another study did not affect lying time but influenced lying bout frequency (Sutherland et al., 2017).

In addition to increasing the softness of the resting area, bedding can promote thermal comfort, particularly in cooler ambient conditions. Pre-weaned calves, relative to adult cattle, have a higher surface area to volume ratio and generate less heat through ruminal fermentation, and therefore expend energy to stay warm at higher ambient temperature thresholds than older cattle (e.g. 10°C for newborns and 0°C for one-month-old calves; Nordlund and Halbach, 2019). In winter, providing calves with a high plane of nutrition, along with resources such as deep bedding to allow for nesting behavior, can help calves avoid unnecessary energy expenditure to stay warm (Nordlund and Halbach, 2019). Lago et al. (2006) reported an inverse relationship between nesting score, describing the ability of the calf to nestle into the bedding provided, and the prevalence of respiratory disease. In cold weather, calves show lying postures with their legs tucked beneath their bodies in winter (air temperature ≤3°C; Gonzalez-Jimenez and Blaxter, 1962; Brunsvold et al., 1985), likely to reduce heat loss. Calves also prefer to increase proximity to a heat lamp during coldest periods of the day (Borderas et al., 2009).

Conversely, heat stress can also be a concern for young calves. A growing body of research has demonstrated that heat stress experienced in utero can affect dairy calf development, health, and adult productivity (reviewed by Dahl et al., 2019). However, very little research has focused on direct effects of heat stress and heat abatement for calves. While some attempt has been made to characterize temperature-humidity index thresholds for heat stress in calves based on physiological measurements (Kovács et al., 2020) limitations in this area of research to date include a lack of behavioral evaluation or consideration of individual variability.

Limited research to date has explored effects of shade provision and fans as cooling mechanisms for pre-weaned calves. For calves housed in outdoor plastic hutches, although the hutch provides shelter from precipitation, supplemental shade from trees or shade structures is needed to prevent solar gain or a greenhouse effect. Older studies have found benefits of supplemental shade structures over plastic hutches in terms of physiological and growth performance measures (Coleman et al., 1996; Spain and Spiers, 1996). As well, some studies have evaluated the effects of ventilation on physiological and growth responses for calves in outdoor hutches (Moore et al., 2012) or cooled with fans indoors (Hill et al., 2011). More research is warranted on cooling calves housed both indoors and in hutches using fans, natural airflow, water, or the combination, to investigate effects on both behavior and physiology and to evaluate how these interventions interact with effects of heat stress experienced in utero.

4.2 Rest and comfort in weaned heifers

Pregnant heifers are motivated to spend more than half the day lying down (12-13 h/d; Jensen et al., 2005), and when adult cows were presented with a tradeoff between feeding and lying, they prioritized lying (Metz, 1985; Munksgaard et al., 2005). Environmental characteristics have been shown to affect lying time and behavioral responses in cattle of all ages. Because cattle are highly motivated to lie down, they will use surfaces they do not prefer or even find aversive, but will spend less time lying down or show hesitation before doing so (e.g. Krohn and Munksgaard, 1993).

Weaned heifers are commonly housed in open-lot or bedded-pack housing systems (e.g. >60% in the United States; USDA, 2016). Housing and bedding conditions in these systems have important implications for rest and comfort. When pregnant heifers were kept in covered housing on dirt floors with higher moisture content, lying time was reduced by 54% (-7 h) relative to dry conditions (67% vs. 90% soil dry matter; Chen et al., 2017). Similar reductions in lying time were found in studies on muddy open lots or pastures where soil dry matter was not quantified (by 3.6-8.6 h/d in Muller et al., 1996 and by 5.7-6.9 h/d in Fisher et al., 2003). Although lying time increased somewhat on subsequent days of exposure, this was likely out of necessity due to exhaustion, rather than acceptance of the wet lying surface. In addition to the reduction in lying time, heifers avoided lying directly on concrete in dry conditions, but chose to spend the majority of their lying time (≥87%; Chen et al., 2017) on this normally aversive, hard surface in muddy conditions. Together, these findings indicate the importance of providing a dry lying surface to ensure heifers have adequate opportunities to lie down.

As they age and eventually enter the lactating herd, most dairy heifers experience a transition to a freestall barn. For example, in the United States, 15% of weaned dairy heifers are housed in freestalls, which increases to 56% for lactating cows (USDA, 2010). In this system, the designated lying area comprises stalls to orient cattle so they eliminate into an alleyway where manure can be easily cleared away while reducing soiling of the bedding surfaces. This strategy is used to keep both bedding and cattle cleaner than in bedded-pack systems, potentially lowering bedding and labor costs and reducing the percentage of dirty cows (10% vs. 22% in freestalls vs. open lots; USDA, 2010).

Although beneficial from a management standpoint, transition to a freestall barn may be stressful for the heifer. Upon first exposure to sand-bedded stalls after being housed on bedded packs, total lying time was reduced by 20%-29% (3-4 h/d; von Kesyerlingk et al., 2011). In addition, undesirable lying behaviors increased. First, heifers spent more time lying in the alley (O'Connell et al., 1993; Kjæstad and Myren, 2001a; von Keyserlingk et al., 2011), where their bodies directly contact manure, which can increase the risk of developing mastitis (Kjæstad and Simensen, 2001). This pattern across studies is interesting given that heifers often avoid lying down on hard or wet surfaces, such as the alley, as described above. Second, they spent more time perching (i.e. standing with the front feet in the stalls and hind feet in the alley; von Keyserlingk et al., 2011), a behavioral risk factor associated with lameness in adult cattle (Galindo and Broom, 2000; Bernardi et al., 2009). Together, these behaviors suggest that, at least initially, heifers may find stalls to be an unintuitive or unappealing place to lie down. Refusal to lie in stalls may occasionally persist into adulthood: 54% of Norwegian dairy producers, out of 184 surveyed, reporting having a least one cow in the herd who lay in the alley (Kjæstad and Myren, 2001b), although stall design and bedding in the adult housing likely play a role.

To date, only a single study has investigated how to ease the adjustment of heifers to freestalls by covering the concrete stall base with mats or by providing feed in the stalls to attract the heifers into entering them (O'Connell et al., 1993). Although no studies have been published yet, there has been recent interest in mounting small brushes on the stalls to attract the heifers or using experienced social models to demonstrate stall use, similar to how older social companions have been shown to increase feeding frequency and duration (De Paula Vieiera et al., 2012). Some commercial dairies have adopted the strategy of introducing heifers to the components of freestalls gradually, making the transition to a stall base without hardware (i.e. loops, neckrail) before moving heifers to a pen with complete stalls; this stepwise approach merits future study. As well, no studies have investigated the best age at which to introduce cattle to freestalls. On some dairies, the first time they encounter freestalls is after the first calving, which likely presents a detrimental combination of stressors. Conversely, the potentially adverse effects of early life exposure to freestall housing and concrete flooring on behavioral expression as well as physical development merit further investigation.

In terms of thermal comfort, a growing body of evidence has evaluated the effects of heat stress on pregnant heifers and cows during their dry period, with negative implications for their immune function and future fertility and production (Ferreira et al., 2016; Fabris et al., 2019). Studies have found that cooling via water soakers and fans can mitigate these negative fitness and production effects (Ferreira et al., 2016), but less work has evaluated the behavioral effects of cooling on heifers. One study on 10-month-old beef steers naïve to water soakers found that they preferred feed bunks with soakers overhead compared to those without, especially in warmer ambient conditions (Parola et al., 2012). In another warm-weather study of pregnant, non-lactating dairy cattle, some of whom were as young as 16 months old, individual use of a pressure-activated shower varied greatly (Legrand et al., 2011), perhaps in part because the shower was not shaded. Together, these studies suggest that younger cattle with limited or no experience with cooling make some voluntary use of heat-abatement resources, and further study on cooling calves and heifers is warranted.

5 Addressing other behavioral needs

Calves seek stimulation and interaction during their diurnal periods of activity, particularly around feeding, and providing resources to accommodate a range of behaviors, such as grooming and play behavior, may be important for calf welfare.

5.1 Grooming behaviour

Cattle commonly exhibit self-grooming and allogrooming under naturalistic conditions (Reinhardt and Reinhardt, 1981). Aside from licking or scratching themselves with a hind leg, self-grooming can be expressed as rubbing or scratching against objects in the environment (Huber et al., 2008) and it has become increasingly common for dairy producers to provide brushes for adult cows to use for grooming. With motivation testing, lactating cows were shown to put forth great effort to gain access to a rotating mechanical brush (McConnachie et al., 2018), and self-grooming rebounds after a period of restraint (Bolinger et al., 1997). Some rotating mechanical brushes are marketed specifically for youngstock (Fig. 4a) and are used for periods of 20-30 min/d in calves ranging in age from 2 weeks to 7 weeks (Zobel et al., 2017; Horvath and

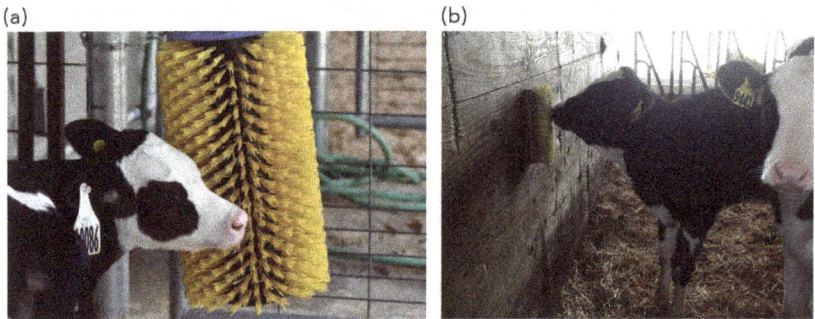

Figure 4 Images of (a) pre-weaned calf contacting a rotating brush (DeLaval mini swing brush) at the University of Florida heifer-raising facility in Alachua, Florida, United States (photo by Catherine Hixson, University of Florida) and (b) a post-weaned heifer investigating a non-rotating brush at the University of Wisconsin heifer-raising facility in Marshfield, Wisconsin, United States (photo by Nancy Esser, Marshfield Agricultural Research Station).

Miller-Cushon, 2019). Characterization of brush use suggests that pre-weaned calves visit them for frequent, short periods (e.g. 2–3 min/visit, 10 times/d; Horvath and Miller-Cushon, 2019).

Recently, researchers have also become interested in providing simple, non-rotating brushes (Fig. 4b) to pre- or post-weaned heifers, which may be a practical and cost-effective way to increase opportunities for grooming behavior. Calves may use these manual brushes for less time than rotating brushes (e.g. pre-weaned calves: 4–16 min/12 h observation period; Pempek et al., 2017; 20-week-old weaned heifers: 21–25 min/24 h on average after the first day of exposure, Van Os et al., 2019), but no research to date has directly compared provision of different grooming resources to cattle of any life stage. Further, there has been no investigation of how many of these objects to provide for group-housed cattle or social dynamics surrounding their use.

Provision of non-rotating brushes may be intended to stimulate grooming behavior, but may also serve to more generally increase environmental complexity or serve as an outlet for oral manipulation (Horvath et al., 2020; Van Os et al., 2019), as shown in Fig. 4b. Likewise, when hemp ropes are mounted in the pen, both 2-week-old dairy calves (Zobel et al., 2017) and older feedlot cattle will regularly chew on them. Recent evidence suggests that provision of manual brushes to pre-weaned calves reduces performance of pen-directed abnormal oral behavior and reduces standing time surrounding feeding (Horvath et al., 2020), suggesting that they may decrease arousal during the period of high activity at feeding.

Expression of allogrooming and self-grooming behavior in young calves and heifers may have a range of welfare implications and also serve as a possible indicator of welfare. The calf is groomed by the dam in the first hours after birth (Jensen, 2011) and a reduction in heart rate of receivers of allogrooming (Laister et al., 2011) may suggest that allogrooming has a calming effect. Expression of both allogrooming and self-grooming may provide an indication of calf welfare from both behavioral and physiological perspectives, as these types of grooming both relate to health. Self-grooming was reduced in calves following a bacterial endotoxin challenge (Borderas et al., 2008). Calves challenged with a respiratory disease pathogen initiated less allogrooming but, interestingly, tended to receive more social grooming from healthy penmates (Hixson et al., 2018).

Environmental factors that affect expression of allogrooming in socially housed calves are not well established, as evidence suggests no effect of group size (Færevik et al., 2007) or brush provision (Horvath and Miller-Cushon, 2019) on expression of allogrooming. However, allogrooming is seen more between familiar penmates (Færevik et al., 2007), indicating its role in social bonding and supporting the importance of maintaining stable social groups. Self-grooming is influenced by environmental factors, such as bedding type (e.g. increasing when bedded on rice hulls or sand vs. long wheat straw; Panivivat et al., 2004).

Some evidence suggests that provision of brushes also stimulates self-grooming in pre-weaned calves, in the form of body-directed licking or scratching, and improves coat cleanliness during weaning (Horvath and Miller-Cushon, 2019).

5.2 Play behaviour

Opportunity for play behavior, both locomotor play (e.g. running and jumping) and social play, may serve an important developmental role for dairy calves. Locomotor play behavior may be affected by the amount of physical space provided to pre-weaned calves (Jensen and Kyhn, 2000). However, effects of space allowance and bedding type on running are not consistently reported; Sutherland et al. (2014) described that calves reared with less space spent more time running in a test arena, whereas behavior in the home pen did not differ. Evidence suggests that bedding material affects play behavior, with more locomotor play (running, head shakes, jumps, kicks, and leaps) observed in calves reared on wood shavings compared to stones (Sutherland et al., 2013).

When play behavior is facilitated after a period of restriction, calves exhibit a rebound effect, suggesting that accommodating locomotor play via social contact, physical space, or a combination of both may be considered a behavioral need. For example, calves reared individually spend more time in locomotor play when introduced to a new space and companions after weaning, compared to calves raised in groups (Valníčková et al., 2015). A rebound effect in locomotor play is also seen when calves reared in smaller space allowances are placed in an open-field test after weaning (Jensen and Kyhn, 2000). However, Bertelsen, and Jensen (2019) found that the duration of social play did not increase following a period of deprivation, suggesting that there is not a motivational build-up for all types of play.

Expression of play behavior may reflect the affective state of calves, as it is often found to be associated with negative states such as hunger and pain. Locomotor play behavior decreases when calves are provided restricted amounts of milk (Jensen et al., 2015) and following disbudding (Mintline et al., 2013; Winder et al., 2017). Given the reduction in play behavior coinciding with these negative states, there has been discussion that play may be considered an indicator of positive emotions; however, evidence in this area is lacking to date (as reviewed by Ahloy-Dallaire et al., 2018).

6 Common themes and developing areas of research

6.1 Environmental complexity and animal welfare

We have addressed specific areas of dairy calf and heifer behavioral needs and corresponding management issues and solutions, yet a broader consideration in intensive housing of livestock is opportunity for behavioral variability and environmental complexity. Ultimately, typical on-farm housing for dairy calves and heifers remains quite restrictive, especially in cases of individual housing for pre-weaned calves, but also in the range of behaviors afforded to group-housed calves. Welfare consequences associated with boredom and frustration in restricted environments are well documented, specifically regarding the development of abnormal behaviors (Mason, 1991). Increasing environmental complexity to accommodate greater behavioral variability may have a range of benefits for behavioral development and welfare, and there are gaps in our knowledge regarding resources and optimal housing environments to provide for developing heifers. Here we summarize some areas of ongoing interest in the impact of environmental complexity on abnormal behaviors, cognition, and boredom.

In dairy calves, non-nutritive oral behaviors are prevalent, and relate to feeding management, including milk feeding with a teat (Section 3.1), and provision of forage in young calves (Section 3.3), as well as weaned heifers (Section 3.4). However, non-nutritive sucking can still develop when calves are provided higher milk allowances. Pen-directed sucking and licking behaviors (Fig. 2) occupy a considerable amount of time, for example, 40-60 min/12 h observation periods (Pempek et al., 2017; Horvath et al. 2020), around 10-20 % of the calf's active time (based on lying time of approximately 18 h/d; Bonk et al., 2013).

Provision of items meant to reduce boredom and abnormal behaviors, such as manual brushes and items to chew on, has not consistently been found to have any effect on these behaviors (Pempek et al., 2017). Non-nutritive oral behaviors (e.g. manipulating substrates, tongue-rolling, and cross-sucking) are also common in veal calves and, although veal calves are subject to a range of different management factors compared with replacement dairy heifers, increased space allowance is one factor which reduces the incidence of abnormal oral behaviors (Leruste et al., 2014). Young calves appear to be highly motivated for stimulation during their diurnal periods of activity, and the approach to reduce development of abnormal oral behaviors may depend on substantially increasing environmental complexity and opportunities for behavioral variability.

It is also well established across species that environmental complexity affects cognitive development. In dairy calves, positive effects of social contact on behavioral flexibility, as assessed in a reversal learning task, has been demonstrated repeatedly (Gaillard et al., 2014; Meagher et al., 2015). Evidence also suggests that increasing feeding complexity, through provision of forage and a teat for sucking, may improve reversal learning ability (Horvath et al., 2017; Horvath and Miller-Cushon, 2020). These early opportunities for behavioral variability may have important implications for lifelong welfare. For example, behavioral flexibility during the pre-weaning period has been associated with the ability to adapt to a novel environment after weaning, where calves with greater success during the reversal learning stage of a procedural learning task subsequently had lower latency to begin consuming feed in a novel pen and exhibited more exploratory behavior (Horvath and Miller-Cushon, 2018). It is well established in other species that environmental complexity affects both cognition and development of abnormal repetitive behaviors, with individuals displaying stereotypic behaviors also having reduced cognitive flexibility (reviewed by Lewis et al., 2006).

Although development of abnormal behaviors is of concern and has been relatively frequently measured and considered in research to date, housing animals in restrictive environments has broader implications for animal welfare. As reviewed by Meagher (2019), boredom is an animal welfare concern associated with restrictive and unvarying environments, evidenced by greater generalized motivation for stimuli in animals in more restrictive housing. A relatively barren housing environment may, in part, explain the contrafreeloading performed by 13-month-old beef heifers (Van Os et al., 2018). Performing work to access a resource that is simultaneously freely available may allow animals to express agency (Špinka, 2019), afford them a sense of control over their environment, or relieve boredom. These mechanisms likewise possibly explain why dairy heifers showed greater excitement, as measured by heart rate and locomotor behaviors (bucking, jumping, kicking), when they completed an operant task to obtain feed compared with receiving the reward freely (Hagen and Broom, 2004). Boredom in dairy calves has yet to receive formal study, but poses welfare concerns as an aversive experience that also has negative health consequences, as evaluated in humans and other animal species (Meagher, 2019).

Inactivity in dairy calves has received some attention, and may reflect a response to unvarying or restrictive environments. Webb et al. (2017) found that inactivity was affected by dietary complexity, examined in calves provided diets varying in restriction of feeding level and supplemental forage. These authors observed some indication of a negative association between duration of inactivity, such as standing or lying idly, and abnormal oral behaviors, including tongue play and object manipulation. This relationship has also been observed in other species, and inactivity is suggested as an alternative response to stereotypic behavior for animals in restrictive housing (Fureix et al., 2016). Individual differences in susceptibility to boredom are also apparent in other species (Meagher, 2019) and bear consideration alongside other individually variable responses seen in dairy calves and heifers.

6.2 Accommodating individual differences and providing choice

In most livestock production systems, animals are managed at a group level, with diets formulated based on average nutritional needs, and many management decisions made using age-based protocols, rather than accommodating individual variability. Whereas much research in this area has historically reported group-average behavioral responses, there is growing interest in understanding causes and consequences of individual variability in behavioral traits.

Research in recent years has explored personality traits in dairy calves, a topic which has long been studied in humans but is of increasing interest in animals, and has important implications for individual welfare (reviewed by Richter and Hintze, 2019). In dairy calves, research in this area has revealed associations between personality traits and behavioral responses related to feeding and socialization. In Section 2.2, we briefly discuss individual variability in social behavior, alongside evidence that social proximity to penmates is associated with individual personality traits (Lecorps et al., 2018). Individual variability in social behavior suggests that designing housing to accommodate different preferences has important welfare implications. In Section 3.2 we discuss evidence that personality traits are associated with initiation of solid feed intake (Neave et al., 2018), an individual metric that is widely variable, as well as research exploring weaning dairy calves based on starter intake rather than age. Evidence that starter intake-based weaning protocols reveal a wide age range at time of weaning (Benetton et al., 2019) suggests the importance of accommodating individual variability in implementing this transition.

Accommodating individual differences in expression of certain behaviors may require increased environmental complexity to provide groups of animals with more choices. Beyond accommodating individual differences, animals may specifically benefit from choices in their environment. In Section 3.3, we discuss evidence that young ruminants may prefer varied feed sources (Meagher et al., 2017). Promoting animal agency and providing choices in environmental resources in livestock production settings have been discussed as important means to improve animal welfare and accommodate the needs of individual animals (Špinka, 2019). This relates to our discussion of complexity of social housing in Section 2.2, where we discuss recent findings that calves altered the use of a visual barrier after disbudding. This developing area of research has the potential to improve individual animal experiences, reducing aversive states such as boredom and facilitating a broader range of motivated behaviors, and accommodating individual preferences in ways that may improve growth performance as well as welfare. Further research to apply these concepts to practical housing solutions for dairy calves and heifers will be beneficial.

7 Conclusion

This chapter provides an overview of key concepts related to accommodating behavioral needs of dairy calves and heifers to improve welfare, including social, nutritional, rest and comfort needs, and expression of other behaviors such as grooming and play. Effects of early experience are seen both in short-term investigation of calf development, and in longer-term behavior of heifers after weaning, emphasizing the importance of refining early-life management strategies. Although the focus here has been on behavioral needs and preferences of dairy calves, evidence throughout supports the link between improved growth performance when behavioral needs are met. Areas of developing research and common themes emerging in recent research include generally addressing environmental complexity and understanding and accommodating individual differences. Research towards improving dairy calf and heifer management is a key component of dairy sustainability, amid ongoing societal concern for animal welfare.

8 Where to look for further information

The following textbook chapter provides a useful overview of similar topics related to calf housing and welfare, with a particular focus on health and early life management, including colostrum management:

- Miller-Cushon, E. K., Leslie, K. E. and DeVries, T. J. 2017. Ensuring the health and welfare of dairy calves and heifers. Chapter 6 in Achieving sustainable production of milk, Volume 3: Dairy herd management and welfare. Burleigh Dodds Science Publishing, Cambridge, UK.

The following review articles provide good overviews of key subjects discussed within this chapter, including effects of social housing for dairy calves, milk allowance, and environmental enrichment in general:

- Costa, J. H. C., von Keyserlingk, M. A. G. and Weary, D. M. 2016. Invited review: Effects of group housing of dairy calves on behavior, cognition, performance, and health. J. Dairy Sci. 99:2453-2467.
- Khan, M. A., Weary, D. M. and von Keyserlingk, M. A. G. 2011. Invited review: Effects of milk ration on solid feed intake, weaning, and performance in dairy heifers. J. Dairy Sci. 94:1071-1081.
- Mandel, R., Whay, H. R., Klement, E. and Nicol, C. J. 2016. Invited review: Environmental enrichment of dairy cows and calves in indoor housing. J. Dairy Sci. 99:1695-1715.

9 References

Ahloy-Dallaire, J., Espinosa, J. and Mason, G. 2018. Play and optimal welfare: does play indicate the presence of positive affective states? *Behav. Processes* 156:3-15.

Appleby, M. C., Weary, D. M. and Chua, B. 2001. Performance and feeding behaviour of calves on ad libitum milk from artificial teats. *Appl. Anim. Behav. Sci.* 74(3): 191-201.

Beauchemin, K. A., Eriksen, L., Nørgaard, P. and Rode, L. M. 2008. Short communication: salivary secretion during meals in lactating dairy cattle. *J. Dairy Sci.* 91(5):2077-2081.

Benetton, J. B., Neave, H. W., Costa, J. H. C., von Keyserlingk, M. A. G. and Weary, D. M. 2019. Automatic weaning based on individual solid feed intake: Effects on behavior and performance of dairy calves. *J. Dairy Sci.* 102:5475-5491.

Bernardi, F., Fregonesi, J. A., Winckler, C., Veira, D. M., von Keyserlingk, M. A. G. and Weary, D. M. 2009. The stall-design paradox: neck rails increase lameness but improve udder and stall hygiene. *J. Dairy Sci.* 92(7):3074-3080.

Bertelsen, M. and Jensen, M. B. 2019. Does dairy calves' motivation for social play behaviour build up over time? *Appl. Anim. Behav. Sci.* 214:18-24.

Bolinger, D. J., Albright, J. L., Morrow-Tesch, J., Kenyon, S. J. and Cunningham, M. D. 1997. The effects of restraint using self-locking stanchions on dairy cows in relation to behavior, feed intake, physiological parameters, health, and milk yield. *J. Dairy Sci.* 80(10):2411-2417.

Bonk, S., Burfeind, O., Suthar, V. S. and Heuwieser, W. 2013. Technical note: evaluation of data loggers for measuring lying behavior in dairy calves. *J. Dairy Sci.* 96(5):3265-3271.

Borderas, T. F., De Passillé, A. M. and Rushen, J. 2008. Behavior of dairy calves after a low dose of bacterial endotoxin. *J. Anim. Sci.* 86(11):2920-2927.

Borderas, T. F., De Passillé, A. M. B. and Rushen, J. 2009. Temperature preferences and feed level of the newborn dairy calf. *Appl. Anim. Behav. Sci.* 120(1-2):56-61.

Brunsvold, R. E., Cramer, C.O. and Larsen, H. J. 1985. Behavior of dairy calves reared in hutches as affected by temperature. *Trans. ASAE* 28(4):1265-1268.

Buchli, C., Raselli, A., Bruckmaier, R. and Hillmann, E. 2017. Contact with cows during the young age increases social competence and lowers the cardiac stress reaction in dairy calves. *Appl. Anim. Behav. Sci.* 187:1-7.

Bučková, K., Špinka, M. and Hintze, S. 2019. Pair housing makes calves more optimistic. *Sci. Rep.* 9(1):20246.

Camiloti, T. V., Fregonesi, J. A., von Keyserlingk, M. A. G. and Weary, D. M. 2012. Short communication: effects of bedding quality on the lying behavior of dairy calves. *J. Dairy Sci.* 95(6):3380-3383.

Catanese, F., Obelar, M., Villalba, J. J. and Distel, R. A. 2013. The importance of diet choice on stress-related responses by lambs. *Appl. Anim. Behav. Sci.* 148(1-2):37-45.

Chen, J. M., Stull, C. L., Ledgerwood, D. N. and Tucker, C. B. 2017. Muddy conditions reduce hygiene and lying time in dairy cattle and increase time spent on concrete. *J. Dairy Sci.* 100(3):2090-2103.

Chua, B., Coenen, E., van, D. J. and Weary, D. M. 2002. Effects of pair versus individual housing on the behavior and performance of dairy calves. *J. Dairy Sci.* 85(2):360-364.

Coleman, D. A., Moss, B. R. and McCaskey, T. A. 1996. Supplemental shade for dairy calves reared in commercial calf hutches in a Southern climate. *J. Dairy Sci.* 79(11):2038-2043.

Costa, J. H. C., Daros, R. R., von Keyserlingk, M. A. G. and Weary, D. M. 2014. Complex social housing reduces food neophobia in dairy calves. *J. Dairy Sci.* 97(12):7804-7810.

Costa, J. H. C., von Keyserlingk, M. A. G. and Weary, D. M. 2016a. Invited review: effects of group housing of dairy calves on behavior, cognition, performance, and health. *J. Dairy Sci.* 99(4):2453-2467.

Costa, J. H. C., Adderley, N. A., Weary, D. M. and von Keyserlingk, M. A. G. 2016b. Short communication: effect of diet changes on sorting behavior of weaned dairy calves. *J. Dairy Sci.* 99(7):5635-5639.

Dado, R. G. and Allen, M. S. 1994. Variation in and relationships among feeding, chewing, and drinking variables for lactating dairy cows. *J. Dairy Sci.* 77(1):132-144.

Dahl, G. E., Skibiel, A. L. and Laporta, J. 2019. In utero heat stress programs reduced performance and health in calves. *Vet. Clin. North Am. Food Anim. Pract.* 35(2):343-353.

de Passillé, A. M. 2001. Sucking motivation and related problems in calves. *Appl. Anim. Behav. Sci.* 72(3):175-187.

de Passillé, A. M., Christopherson, R. and Rushen, J. 1993. Non-nutritive sucking by the calf and postprandial secretion of insulin, CCK and gastrin. *Physiol. Behav.* 54(6):1069-1073.

de Passillé, A. M., Sweeney, B. and Rushen, J. 2010. Cross-sucking and gradual weaning of dairy calves. *Appl. Anim. Behav. Sci.* 124(1-2):11-15.

de Passillé, A. M. and Rushen, J. 2016. Using automated feeders to wean calves fed large amounts of milk according to their ability to eat solid feed. *J. Dairy Sci.* 99(5):3578-3583.

De Paula Vieira, A., Guesdon, V., de Passille, A. M., von Keyserlingk, M. A. G. and Weary, D. M. 2008. Behavioral indicators of hunger in dairy calves. *Appl. Anim. Behav. Sci.* 109(2-4):180-189.

De Paula Vieira, A., von Keyserlingk, M. A. G. and Weary, D. M. 2010. Effects of pair versus single housing on performance and behavior of dairy calves before and after weaning from milk. *J. Dairy Sci.* 93(7):3079-3085.

De Paula Vieira, A., von Keyserlingk, M. A. G. and Weary, D. M. 2012. Presence of an older weaned companion influences feeding behavior and improves performance of dairy calves before and after weaning from milk. *J. Dairy Sci.* 95(6):3218-3224.

DeVries, T. J., Dohme, F. and Beauchemin, K. A. 2008. Repeated ruminal acidosis challenges in lactating dairy cows at high and low risk for developing acidosis: feed sorting. *J. Dairy Sci.* 91(10):3958-3967.

DeVries, T. J., Schwaiger, T., Beauchemin, K. A. and Penner, G. B. 2014. Impact of severity of ruminal acidosis on feed-sorting behaviour of beef cattle. *Anim. Prod. Sci.* 54(9):1238-1242.

Duve, L. R., Weary, D. M., Halekoh, U. and Jensen, M. B. 2012. The effects of social contact and milk allowance on responses to handling, play, and social behavior in young dairy calves. *J. Dairy Sci.* 95(11):6571-6581.

Eckert, E., Brown, H. E., Leslie, K. E., DeVries, T. J. and Steele, M. A. 2015. Weaning age affects growth, feed intake, gastrointestinal development, and behavior in Holstein calves fed an elevated plane of nutrition during the preweaning stage. *J. Dairy Sci.* 98(9):6315-6326.

Ede, T., von Keyserlingk, M. A. G. and Weary, D. M. 2018. Approach-aversion in calves following injections. *Sci. Rep.* 8(1):9443.

Fabris, T. F., Laporta, J., Skibiel, A. L., Corra, F. N., Sen, B. D., Wohlgemuth, S. E. and Dahl, G. E. 2019. Effect of heat stress during early, late, and entire dry period on dairy cattle. *J. Dairy Sci.* 102(6):5647-5656.

Færevik, G., Jensen, M. B. and Bøe, K. E. 2006. Dairy calves social preference and the significance of a companion animal during separation from the group. *Appl. Anim. Behav. Sci.* 99(3-4):205--221.

Færevik, G., Andersen, I. L., Jensen, M. B. and Bøe, K. E. 2007. Increased group size reduces conflicts and strengthens the preference for familiar group mates after regrouping of weaned dairy calves (Bos taurus). *Appl. Anim. Behav. Sci.* 108(3-4):215-228.

Favreau-Peigné, A., Baumont, R. and Ginane, C. 2013. Food sensory characteristics: their unconsidered roles in the feeding behavior of domestic ruminants. *Animal* 7(5):806-813.

Ferreira, F. C., Gennari, R. S., Dahl, G. E. and De Vries, A. 2016. Economic feasibility of cooling dry cows across the United States. *J. Dairy Sci.* 99(12):9931-9941.

Fisher, A. D., Stewart, M., Verkerk, G. A., Morrow, C. J. and Matthews, L. R. 2003. The effects of surface type on lying behaviour and stress responses of dairy cows during periodic weather-induced removal from pasture. *Appl. Anim. Behav. Sci.* 81(1):1-11.

Flower, F. C. and Weary, D. M. 2001. Effects of early separation on the dairy cow and calf: 2. Separation at 1 day and 2 weeks after birth. *Appl. Anim. Behav. Sci.* 70(4):275-284.

Fröberg, S., Aspegren-Guldorff, A., Olsson, I., Marin, B., Berg, C., Hernandez, C., Galina, C. S., Lidfors, L. and Svennersten-Sjaunja, K. 2007. Effect of restricted suckling on milk yield, milk composition and udder health in cows and behaviour and weight gain in calves, in dual-purpose cattle in the tropics. *Trop. Anim. Health Prod.* 39(1):71-81.

Fureix, C., Walker, M., Harper, L., Reynolds, K., Saldivia-Woo, A. and Mason, G. 2016. Stereotypic behaviour in standard non-enriched cages is alternative to depression-like responses in C57BL/6 mice. *Behav. Brain Res.* 305:186-190.

Gaillard, C., Meagher, R. K., von Keyserlingk, M. A. G. and Weary, D. M. 2014. Social housing improves dairy calves' performance in two cognitive tests. *PLoS ONE* 9(2):e90205.

Galindo, F. and Broom, D. M. 2000. The relationships between social behaviour of dairy cows and the occurrence of lameness in three herds. *Res. Vet. Sci.* 69(1):75-79.

Gingerich, K. N., V. Choulet and Miller-Cushon, E. K. 2020. Disbudding affects use of a shelter provided to group-housed dairy calves. *J. Dairy Sci.* https://doi.org/10.3168/jds.2020-18267.

Gonzalez-Jimenez, E. and Baxter, K. L. 1962. The metabolism and thermal regulation of calves in the first month of life. *Br. J. Nutr.* 16:199-212.

Greter, A. M., Miller-Cushon, E. K., McBride, B. W., Widowski, T. M., Duffield, T. F. and DeVries, T. J. 2015. Short communication: limit feeding affects behavior patterns and feed motivation of dairy heifers. *J. Dairy Sci.* 98(2):1248-1254.

Hagen, K. and Broom, D. M. 2004. Emotional reactions to learning in cattle. *Appl. Anim. Behav. Sci.* 85(3-4):203-213.

Haley, D. B., Rushen, J., Duncan, I. J., Widowski, T. M. and De Passillé, A. M. 1998. Effects of resistance to milk flow and the provision of hay on nonnutritive sucking by dairy calves. *J. Dairy Sci.* 81(8):2165-2172.

Hänninen, L., de Passillé, A. M. and Rushen, J. 2005. The effect of flooring type and social grouping on the rest and growth of dairy calves. *Appl. Anim. Behav. Sci.* 91(3-4):193-204.

Hill, T. M., Bateman, H. G., Aldrich, J. M. and Schlotterbeck, R. L. 2011. Comparisons of housing, bedding, and cooling options for dairy calves. *J. Dairy Sci.* 94(4):2138-2146.

Hixson, C. L., Krawczel, P. D., Caldwell, J. M. and Miller-Cushon, E. K. 2018. Behavioral changes in group-housed dairy calves infected with Mannheimia haemolytica. *J. Dairy Sci.* 101(11):10351-10360.

Holm, L., Jensen, M. B. and Jeppesen, L. L. 2002. Calves' motivation for access to two different types of social contact measured by operant conditioning. *Appl. Anim. Behav. Sci.* 79(3):175-194.

Horvath, K. C., Fernandez, M. and Miller-Cushon, E. K. 2017. The effect of feeding enrichment in the milk-feeding stage on the cognition of dairy calves in a T-maze. *Appl. Anim. Behav. Sci.* 187:8-14.

Horvath, K. C. and Miller-Cushon, E. K. 2017. The effect of milk-feeding method and hay provision on the development of feeding behavior and non-nutritive oral behavior of dairy calves. *J. Dairy Sci.* 100(5):3949-3957.

Horvath, K. C. and Miller-Cushon, E. K. 2018. Characterizing social behavior, activity, and associations between cognition and behavior upon social grouping of weaned dairy calves. *J. Dairy Sci.* 101(8):7287-7296.

Horvath, K. C. and Miller-Cushon, E. K. 2019. Characterizing grooming behavior patterns and the influence of brush access on the behavior of group-housed dairy calves. *J. Dairy Sci.* 102(4):3421-3430.

Horvath, K. C. and Miller-Cushon, E. K. 2020. Effects of hay provision and presentation on cognitive development in dairy calves. *PLoS ONE* 15(9): e0238038. https://doi.org /10.1371/journal.pone.0238038.

Horvath, K. C., Allen, A. N. and Miller-Cushon, E. K. 2020. Effects of access to stationary brushes and chopped hay on behavior and performance of individually housed dairy calves. *J. Dairy Sci.* 103:8421-8432.

Hosseinkhani, A., DeVries, T. J., Proudfoot, K. L., Valizadeh, R., Veira, D. M. and von Keyserlingk, M. A. G. 2008. The effects of feed bunk competition on the feed sorting behavior of close-up dry cows. *J. Dairy Sci.* 91(3): 1115-1121.

Huber, R., Baumung, R., Wurzinger, M., Semambo, D., Mwai, O. and Winckler, C. 2008. Grazing, social and comfort behaviour of Ankole and crossbred (Ankole x Holstein) heifers on pasture in south western Uganda. *Appl. Anim. Behav. Sci.* 112(3-4): 223-234.

Jasper, J. and Weary, D. M. 2002. Effects of ad libitum milk intake on dairy calves. *J. Dairy Sci.* 85(11):3054-3058.

Jensen, M. B., Vestergaard, K. S., Krohn, C. C. and Munksgaard, L. 1997. Effect of single versus group housing and space allowance on responses of calves during open-field tests. *Appl. Anim. Behav. Sci.* 54(2-3):109-121.

Jensen, M. B. and Kyhn, R. 2000. Play behaviour in group-housed dairy calves, the effect of space allowance. *Appl. Anim. Behav. Sci.* 67(1-2):35-46.

Jensen, M. B., Pedersen, L. J. and Munksgaard, L. 2005. The effect of reward duration on demand functions for rest in dairy heifers and lying requirements as measured by demand functions. *Appl. Anim. Behav. Sci.* 90(3-4):207-217.

Jensen, M. B. and Budde, M. 2006. The effects of milk feeding method and group size on feeding behavior and cross-sucking in group-housed dairy calves. *J. Dairy Sci.* 89(12):4778-4783.

Jensen, M. B., de Passillé, A. M., von Keyserlingk, M. A. G. and Rushen, J. 2008. A barrier can reduce competition over teats in pair-housed milk-fed calves. *J. Dairy Sci.* 91(4):1607-1613.

Jensen, M. B. 2011. The early behaviour of cow and calf in an individual calving pen. *Appl. Anim. Behav. Sci.* 134(3-4):92-99.

Jensen, M. B., Duve, L. R. and Weary, D. M. 2015. Pair housing and enhanced milk allowance increase play behavior and improve performance in dairy calves. *J. Dairy Sci.* 98(4):2568-2575.

Johnsen, J. F., de Passille, A. M., Mejdell, C. M., Bøe, K. E., Grøndahl, A. M., Beaver, A., Rushen, J. and Weary, D. M. 2015. The effect of nursing on the cow-calf bond. *Appl. Anim. Behav. Sci.* 163:50-57.

Johnsen, J. F., Zipp, K. A., Kälber, T., de Passillé, A. Md, Knierim, U., Barth, K. and Mejdell, C. M. 2016. Is rearing calves with the dam a feasible option for dairy farms?–current and future research. *Appl. Anim. Behav. Sci.* 181:1-11.

Jorgensen, M. W., Adams-Progar, A., de Passillé, A. M., Rushen, J., Godden, S. M., Chester-Jones, H. and Endres, M. I. 2017. Factors associated with dairy calf health in automated feeding systems in the Upper Midwest United States. *J. Dairy Sci.* 100(7):5675-5686.

Jung, J. and Lidfors, L. 2001. Effects of amount of milk, milk flow and access to a rubber teat on cross-sucking and non-nutritive sucking in dairy calves. *Appl. Anim. Behav. Sci.* 72(3):201-213.

Keil, N. M. and Langhans, W. 2001. The development of intersucking in dairy calves around weaning. *Appl. Anim. Behav. Sci.* 72(4):295-308.

Keunen, J. E., Plaizier, J. C., Kyriazakis, L., Duffield, T. F., Widowski, T. M., Lindinger, M. I. and McBride, B. W. 2002. Effects of a subacute ruminal acidosis model on the diet selection of dairy cows. *J. Dairy Sci.* 85(12):3304-3313.

Khan, M. A., Lee, H. J., Lee, W. S., Kim, H. S., Ki, K. S., Hur, T. Y., Suh, G. H., Kang, S. J. and Choi, Y. J. 2007. Structural growth, rumen development, and metabolic and immune responses of Holstein male calves fed milk through step-down and conventional methods. *J. Dairy Sci.* 90(7):3376-3387.

Khan, M. A., Weary, D. M. and von Keyserlingk, M. A. G. 2011a. Invited review: effects of milk ration on solid feed intake, weaning, and performance in dairy heifers. *J. Dairy Sci.* 94(3):1071-1081.

Khan, M. A., Weary, D. M. and von Keyserlingk, M. A. G. 2011b. Hay intake improves performance and rumen development of calves fed higher quantities of milk. *J. Dairy Sci.* 94(7):3547-3553.

Kilgour, R. J. 2012. In pursuit of "normal": a review of the behaviour of cattle at pasture. *Appl. Anim. Behav. Sci.* 138(1-2):1-11.

Kitts, B. L., Duncan, I. J. H., McBride, B. W. and DeVries, T. J. 2011. Effect of the provision of a low-nutritive feedstuff on the behavior of dairy heifers fed a high-concentrate ration in a limited amount. *J. Dairy Sci.* 94(2):940-950.

Kjæstad, H. P. and Myren, H. J. 2001a. Failure to use cubicles and concentrate dispenser by heifers after transfer from rearing accommodation to milking herd. *Acta Vet. Scand.* 42(1):171-180.

Kjæstad, H. P. and Myren, H. J. 2001b. Cubicle refusal in Norwegian dairy herds. *Acta Vet. Scand.* 42(1):181-187.

Kjæstad, H. P. and Simensen, E. 2001. Cubicle refusal and rearing accommodation as possible mastitis risk factors in cubicle-housed dairy heifers. *Acta Vet. Scand.* 42(1):123-130.

Kovács, L., Kézér, F. L., Póti, P., Boros, N. and Nagy, K. 2020. Short communication: upper critical temperature-humidity index for dairy calves based on physiological stress variables. *J. Dairy Sci.* 103(3):2707-2710.

Krohn, C. C. and Munksgaard, L. 1993. Behaviour of dairy cows kept in extensive (loose housing/pasture) or intensive (tie stall) environments II. Lying and lying-down behavior. *Appl. Anim. Behav. Sci.* 37(1):1-16.

Kyriazakis, I. and Oldham, J. D. 1993. Diet selection in sheep: the ability of growing lambs to select a diet that meets their crude protein (nitrogen x 6.25) requirements. *Br. J. Nutr.* 69(3):617-629.

Lago, A., McGuirk, S. M., Bennet, T. B., Cook, N. B. and Nordlund, K. V. 2006. Calf respiratory disease and pen microenvironments in naturally ventilated calf barns in winter. *J. Dairy Sci.* 89:4014-4025.

Laister, S., Stockinger, B., Regner, A. M., Zenger, K., Knierim, U. and Winckler, C. 2011. Social licking in dairy cattle—effects on heart rate in performers and receivers. *Appl. Anim. Behav. Sci.* 130(3-4):81-90.

Lecorps, B., Kappel, S., Weary, D. M. and von Keyserlingk, M. A. G. 2018. Dairy calves' personality traits predict social proximity and response to an emotional challenge. *Sci. Rep.* 8(1):16350.

Legrand, A., Schütz, K. E. and Tucker, C. B. 2011. Using water to cool cattle: behavioral and physiological changes associated with voluntary use of cow showers. *J. Dairy Sci.* 94(7):3376-3386.

Lewis, M. H., Tanimura, Y., Lee, L. W. and Bodfish, J. W. 2006. Animal models of restricted repetitive behavior in autism. *Behav. Brain Res.* 10:66-74.

Lidfors, L. M. and Isberg, L. 2003. Intersucking in dairy cattle – review and questionnaire. *Appl. Anim. Behav. Sci.* 80(3):207-231.

Lidfors, L. M., Moran, D., Jung, J., Jensen, P. and Castren, H. 1994. Behaviour at calving and choice of calving place in cattle kept in different environments. *Appl. Anim. Behav. Sci.* 42(1):11-28.

Lindstrom, T. and Redbo, I. 2000. Effect of feeding duration and rumen fill on behaviour in dairy cows. *Appl. Anim. Behav. Sci.* 70(2):83-97.

Leruste, H., Brscic, M., Cozzi, G., Kemp, B., Wolthuis-Fillerup, M., Lensink, B. J., Bokkers, E. A. M. and van Reenen, C. G. 2014. Prevalence and potential influencing factors of non-nutritive oral behaviors of veal calves on commercial farms. *J. Dairy Sci.* 97(11):7021-7030.

Loberg, J. and Lidfors, L. 2001. Effect of milkflow rate and presence of a floating nipple on abnormal sucking between dairy calves. *Appl. Anim. Behav. Sci.* 72(3): 189-199.

Marcé, C., Guatteo, R., Bareille, N. and Fourichon, C. 2010. Dairy calf housing systems across Europe and risk for calf infectious diseases. *Animal* 4(9):1588-1596.

Mason, G. J. 1991. Stereotypies: a critical review. *Anim. Behav.* 41(6):1015-1037.

McConnachie, E., Smid, A. M. C., Thompson, A. J., Weary, D. M., Gaworski, M. A. and von Keyserlingk, M. A. G. 2018. Cows are highly motivated to access a grooming substrate. *Biol. Lett.* 14(8):pii:20180303.

Meagher, R. K., Daros, R. R., Costa, J. H. C., von Keyserlingk, M. A. G., Hötzel, M. J. and Weary, D. M. 2015. Effects of degree and timing of social housing on reversal learning and response to novel objects in dairy calves. *PLoS ONE* 10(8):e0132828.

Meagher, R. K., Weary, D. M. and von Keyserlingk, M. A. G. 2017. Some like it varied: individual differences in preference for feed variety in dairy heifers. *Appl. Anim. Behav. Sci.* 195:8-14.

Meagher, R. K. 2019. Is boredom an animal welfare concern? *Animal Welfare* 28:21-32.

Meagher, R. K., Beaver, A., Weary, D. M. and von Keyserlingk, M. A. G. 2019. Invited review: a systematic review of the effects of prolonged cow–calf contact on behavior, welfare, and productivity. J. Dairy Sci. 102(7):5765-5783.

Metz, J. H. M. 1985. The reaction of cows to a short-term deprivation of lying. Appl. Anim. Behav. Sci. 13(4):301-307.

Miller-Cushon, E. K., Bergeron, R., Leslie, K. E. and DeVries, T. J. 2013a. Effect of milk feeding level on development of feeding behavior in dairy calves. J. Dairy Sci. 96(1):551-564.

Miller-Cushon, E. K., Bergeron, R., Leslie, K. E., Mason, G. J. and DeVries, T. J. 2013b. Effect of early exposure to different feed presentations on feed sorting of dairy calves. J. Dairy Sci. 96(7):4624-4633.

Miller-Cushon, E. K., Bergeron, R., Leslie, K. E., Mason, G. J. and DeVries, T. J. 2014a. Competition during the milk-feeding stage influences the development of feeding behavior of pair-housed dairy calves. J. Dairy Sci. 97(10):6450-6462.

Miller-Cushon, E. K., Montoro, C., Ipharraguerre, I. R. and Bach, A. 2014b. Dietary preference in dairy calves for feed ingredients high in energy and protein. J. Dairy Sci. 97(3):1634-1644.

Miller-Cushon, E. K., Terré, M., DeVries, T. J. and Bach, A. 2014c. The effect of palatability of protein source on dietary selection in dairy calves. J. Dairy Sci. 97(7):4444-4454.

Miller-Cushon, E. K. and DeVries, T. J. 2015. Invited review: development and expression of dairy calf feeding behaviour. Can. J. Anim. Sci. 95(3):341-350.

Miller-Cushon, E. K. and DeVries, T. J. 2016. Effect of social housing on the development of feeding behavior and social feeding preferences of dairy calves. J. Dairy Sci. 99(2):1406-1417.

Mintline, E. M., Stewart, M., Rogers, A. R., Cox, N. R., Verkerk, G. A., Stookey, J. M., Webster, J. R. and Tucker, C. B. 2013. Play behavior as an indicator of animal welfare: disbudding in dairy calves. Appl. Anim. Behav. Sci. 144(1-2):22-30.

Mirzaei, M., Dadkhah, N., Baghbanzadeh-Nobari, B., Agha-Tehrani, A., Eshraghi, M., Imani, M., Shiasi-Sardoabi, R. and Ghaffari, M. H. 2018. Effects of preweaning total plane of milk intake and weaning age on intake, growth performance, and blood metabolites of dairy calves. J. Dairy Sci. 101(5):4212-4220.

Montoro, C., Ipharraguerre, I. and Bach, A. 2011. Effect of flavoring a starter in a same manner as a milk replacer on intake and performance of calves. Anim. Feed Sci. Tech. 164(1-2):130-134.

Montoro, C. and Bach, A. 2012. Voluntary selection of starter feed ingredients offered separately to nursing calves. Livest. Sci. 149(1-2):62-69.

Moore, D. A., Duprau, J. L. and Wenz, J. R. 2012. Short communication: effects of dairy calf hutch elevation on heat reduction, carbon dioxide concentration, air circulation, and respiratory rates. J. Dairy Sci. 95(7):4050-4054.

Muller, C. J. C., Botha, J. A. and Smith, W. A. 1996. Effect of confinement area on production, physiological parameters and behaviour of Friesian cows during winter in a temperate climate. S. Afr. J. Anim. Sci. 26:1-5.

Munksgaard, L., Jensen, M. B., Pedersen, L. J., Hansen, S. W. and Matthews, L. 2005. Quantifying behavioral priorities: effects of time constraints on the behavior of dairy cows, Bos taurus. Appl. Anim. Behav. Sci. 92(1-2):3-14.

Neave, H. W., Costa, J. H. C., Weary, D. M. and von Keyserlingk, M. A. G. 2018. Personality is associated with feeding behavior and performance in dairy calves. J. Dairy Sci. 101(8):7437-7449.

Nicol, A. M. and Sharafeldin, M. A. 1975. Observations on the behaviour of single-suckled calves from birth to 120 days. *Proc. N.Z. Soc. Anim. Prod.* 35:221–230.

Nordlund, K. V. and Halbach, C. E. 2019. Calf barn design to optimize health and ease of management. *Vet. Clin. North Am. Food Anim. Pract.* 35(1):29–45.

O'Connell, J. M., Giller, P. S. and Meaney, W. J. 1993. Weaning training and cubicle usage as heifers. *Anim. Beh.* 37(3):185–195.

Owens, F. N., Secrist, D. S., Hill, W. J. and Gill, D. R. 1998. Acidosis in cattle: a review. *J. Anim. Sci.* 76(1):275–286.

Panivivat, R., Kegley, E. B., Pennington, J. A., Kellogg, D. W. and Krumpelma, S. L. 2004. Growth performance and health of dairy calves bedded with different types of materials. *J. Dairy Sci.* 87(11):3736–3745.

Parola, F., Hillmann, E., Schütz, K. E. and Tucker, C. B. 2012. Preferences for overhead sprinklers by naïve beef steers: test of two nozzle types. *Appl. Anim. Behav. Sci.* 137(1–2):13–22.

Pempek, J. A., Eastridge, M. L. and Proudfoot, K. L. 2017. The effect of a furnished individual hutch pre-weaning on calf behavior, response to novelty, and growth. *J. Dairy Sci.* 100(6):4807–4817.

Proudfoot, K. L., Jensen, M. B., Weary, D. M. and von Keyserlingk, M. A. G. 2014. Dairy cows seek isolation at calving and when ill. *J. Dairy Sci.* 97(5):2731–2739.

Provenza, F. D. and Balph, D. F. 1987. Diet learning by domestic ruminants: theory, evidence and practical implications. *Appl. Anim. Behav. Sci.* 18(3–4):211–232.

Redbo, I. 1990. Changes in duration and frequency of stereotypies and their adjoining behaviours in heifers, before, during and after the grazing period. *Appl. Anim. Behav. Sci.* 26(1–2):57–67.

Redbo, I. and Nordblad, A. 1997. Stereotypies in heifers are affected by feeding regime. *Appl. Anim. Behav. Sci.* 53(3):193–202.

Reinhardt, V. and Reinhardt, A. 1981. Cohesive relationships in a cattle herd (Bos indicus). *Behaviour* 77(3):121–150.

Richter, S. H. and Hintze, S. 2019. From the individual to the population – and back again? Emphasising the role of the individual in animal welfare science. *Appl. Anim. Behav. Sci.* 212:1–8.

Roth, B. A., Hillman, E., Stauffacher, M. and Keil, N. M. 2008. Improved weaning reduces cross-sucking and may improve weight gain in dairy calves. *Appl. Anim. Behav. Sci.* 111(3–4):251–261.

Roth, B. A., Barth, K., Gygax, L. and Hillmann, E. 2009. Influence of artificial vs. mother-bonded rearing on sucking behaviour, health and weight gain in calves. *Appl. Anim. Behav. Sci.* 119(3–4):143–150.

Roura, E., Humphrey, B., Tedo, G. and Ipharraguerre, I. 2008. Unfolding the codes of short-term feed appetence in farm and companion animals. A comparative oronasal nutrient sensing biology review. *Can. J. Anim. Sci.* 88(4):535–558.

Sato, S., Wood-Gush, D. G. M. and G, W. 1987. Observations on creche behaviour in suckler calves. *Behav. Processes* 15(2–3):333–343.

Scott, L. L. and Provenza, F. D. 1998. Variety of foods and flavors affects selection of foraging location by sheep. *Appl. Anim. Behav. Sci.* 61(2): 113–122.

Spain, J. N. and Spiers, D. E. 1996. Effects of supplemental shade on thermoregulatory response of calves to heat challenge in a hutch environment. *J. Dairy Sci.* 79(4):639–646.

Špinka, M. 2019. Animal agency, animal awareness and animal welfare. *Anim. Welf.* 28(1):11-20.

Sutherland, M. A., Stewart, M. and Schütz, K. E. 2013. Effects of two substrate types on the behaviour, cleanliness and thermoregulation of dairy calves. *Appl. Anim. Behav. Sci.* 147(1-2):19-27.

Sutherland, M. A., Worth, G. M., Schütz, K. E. and Stewart, M. 2014. Rearing substrate and space allowance influences locomotor play behaviour of dairy calves in an arena test. *Appl. Anim. Behav. Sci.* 154:8-14.

Sutherland, M. A., Worth, G. M., Cameron, C., Ross, C. M. and Rapp, D. 2017. Health, physiology, and behavior of dairy calves reared on 4 different substrates. *J. Dairy Sci.* 100(3):2148-2156.

Sweeney, B. C., Rushen, J. P., Weary, D. M. and de Passillé, A. M. 2010. Duration of weaning, starter intake, and weight gain of dairy calves fed large amounts of milk. *J. Dairy Sci.* 93(1):148-152.

Thomas, T. J., Weary, D. M. and Appleby, M. C. 2001. Newborn and 5-week-old calves vocalize in response to milk deprivation. *Appl. Anim. Behav. Sci.* 74(3):165-173.

USDA 2010. *Facility Characteristics and Cow Comfort on U.S. Dairy Operations, 2007.* UDSA-Animal and Plant Health Inspection Service-Veterinary Services, Centers for Epidemiology and Animal Health, Fort Collins, CO.

USDA 2016. *Dairy Cattle Management Practices in the United States, 2014.* UDSA-Animal and Plant Health Inspection Service-Veterinary Services, Centers for Epidemiology and Animal Health, Fort Collins, CO.

Valníčková, B., Stěhulová, I., Šárová, R. and Špinka, M. 2015. The effect of age at separation from the dam and presence of social companions on play behavior and weight gain in dairy calves. *J. Dairy Sci.* 98(8):5545-5556.

Van Os, J. M. C., Mintline, E. M., DeVries, T. J. and Tucker, C. B. 2018. Domestic cattle (Bos taurus taurus) are motivated to obtain forage and demonstrate contrafreeloading. *PLoS ONE* 13(3):e0193109.

Van Os, J. M. C., Goldstein, S. A., Weary, D. M. and von Keyserlingk, M. A. G. 2019. Brush use by dairy heifers. Proceedings of the 53rd Congress of the International Society for Applied Ethology, Bergen, Norway.

Vasseur, E., Borderas, F., Cue, R. I., Lefebvre, D., Pellerin, D., Rushen, J., Wade, K. M. and de Passillé, A. M. 2010. A survey of dairy calf management practices in Canada that affect animal welfare. *J. Dairy Sci.* 93(3):1307-1315.

Vaughan, A., Miguel-Pacheco, G. G., de Passillé, A. M. and Rushen, J. 2016. Reciprocated cross sucking between dairy calves after weaning off milk does not appear to negatively affect udder health or production. *J. Dairy Sci.* 99(7):5596-5603.

Ventura, B. A., von Keyserlingk, M. A., Schuppli, C. A. and Weary, D. M. 2013. Views on contentious practices in dairy farming: the case of early cow-calf separation. *J. Dairy Sci.* 96(9):6105-6116.

Veissier, I., de Passillé, A. M., Després, G., Rushen, J., Charpentier, I., Ramirez de la Fe, A. R. and Pradel, P. 2002. Does nutritive and non-nutritive sucking reduce other oral behaviors and stimulate rest in calves? *J. Anim. Sci.* 80(10):2574-2587.

von Keyserlingk, M. A. G., Cunha, G. E., Fregonesi, J. A. and Weary, D. M. 2011. Introducing heifers to freestall housing. *J. Dairy Sci.* 94(4):1900-1907.

Weary, D. M., Jasper, J. and Hötzel, M. J. 2008. Understanding weaning distress. *Appl. Anim. Behav. Sci.* 110(1-2):24-41.

Webb, L. E., Bokkers, E. A. M., Engel, B., Gerrits, W. J. J., Berends, H. and van Reenen, C. G. 2012. Behaviour and welfare of veal calves fed different amounts of solid feed supplemented to a milk replacer ration adjusted for similar growth. *Appl. Anim. Behav. Sci.* 136(2-4):108-116.

Webb, L. E., Engel, B., Berends, H., van Reenen, C. G., Walter, W. J. J., de Boer, I. J. M. and Bokkers, E. A. M. 2014a. What do calves choose to eat and how do preferences affect behaviour? *Appl. Anim. Behav. Sci.* 161:7-19.

Webb, L. E., Jensen, M. B., Engel, B., van Reenen, C. G., Gerrits, W. J. J., de Boer, I. J. M. and Bokkers, E. A. M. 2014b. Chopped or long roughage: what do calves prefer? Using cross point analysis of double demand functions. *PLoS ONE* 9(2):e88778.

Webb, L. E., Engel, B., van Reenen, K. and Bokkers, E. A. M. 2017. Barren diets increase wakeful inactivity in calves. *Appl. Anim. Behav. Sci.* 197:9-14.

Whalin, L., Weary, D. M. and von Keyserlingk, M. A. G. 2018. Short communication: pair housing dairy calves in modified calf hutches. *J. Dairy Sci.* 101(6):5428-5433.

Winder, C. B., LeBlanc, S. J., Haley, D. B., Lissemore, K. D., Godkin, M. A. and Duffield, T. F. 2017. Clinical trial of local anesthetic protocols for acute pain associated with caustic paste disbudding in dairy calves. *J. Dairy Sci.* 100(8):6429-6441.

Wormsbecher, L., Bergeron, R., Haley, D., de Passillé, A. M., Rushen, J. and Vasseur, E. 2017. A method of outdoor housing dairy calves in pairs using individual calf hutches. *J. Dairy Sci.* 100(9):7493-7506.

Worth, G. M., Schütz, K. E., Stewart, M., Cave, V. M., Foster, M. and Sutherland, M. A. 2015. Dairy calves' preference for rearing substrate. *Appl. Anim. Behav. Sci.* 168:1-9.

Zobel, G., Neave, H. W., Henderson, H. V. and Webster, J. R. 2017. Calves use an automated brush and a hanging rope when pair-housed. *Animals* 7(11):84.

Chapter 3

Managing calves/youngstock to optimise dairy herd health

John F. Mee, Teagasc, Ireland

1 Introduction

The focus of this chapter is on how better female calf/youngstock management can improve dairy herd productivity and health; 'the calves of today are the cows of tomorrow'. Youngstock diseases incur both direct costs of prevention and therapy and indirect costs associated with subsequent reduced dairy herd health and performance, a theme that is emphasised here. In addition, calf morbidity and mortality is a welfare issue assuming greater importance now due to increased consumer scrutiny of management practice on modern dairy farms (Mee, 2013).

Here, calves are defined as cattle from birth to weaning (fed colostrum, milk or milk replacer) and replacement heifers as cattle from weaning to first calving; youngstock encompasses both (from birth to calving). Unless specified, the material refers to female Holstein-Friesian cattle managed in temperate dairy industries internationally.

The emphasis is on modifiable risk factors and management practices with a view to improving youngstock immunity and reducing dependence on antimicrobial use (AMU). Although not specifically addressed here, the role of

http://dx.doi.org/10.19103/AS.2020.0086.11

good animal welfare as preventative medicine has recently attracted attention as a drug-free way of improving immune function (Dawkins, 2019). Successful management of youngstock is the start of the lifecycle to optimal dairy herd health (Lorenz et al., 2011). However, farmers are often blind to this axiom.

2 Costs of heifer rearing

One way to convince farmers of the value of good youngstock management is to demonstrate the high, but variable costs of rearing replacement heifers and the effects of management on these costs. It is accepted, however, that farmers are also motivated by the principles of good calf welfare and pride in producing healthy, well-grown heifers as well as by economics. Heifer rearing is the second largest expense on a dairy operation (15–20% of total costs). The average costs of rearing a replacement dairy heifer in recent international calculations, both in confinement- and pasture-based systems vary between €1545 (Shalloo et al., 2014) and €2450 (CAFRE, 2020). However, there can be more than 95% difference in costs between the top and the bottom 25% of farmers (CAFRE, 2020). Most of this extra cost is due to delayed first calving and it takes, on average, two lactations to recoup these costs (Boulton et al., 2017). The rearing costs of a heifer that experienced disease at least once were, on average, €95 higher than those of healthy heifers (Modh Nor et al., 2012). For example, the most recent estimate of the costs of respiratory disease in calves put the short-term costs alone at nearly €40/case (Dubrovsky et al., 2020). With recent international improvements in dairy cow fertility and increasing use of sexed semen, it may be prudent in future to shift away from producing excess heifers for herds that are not expanding and concentrate on producing fewer, higher quality, healthier heifers (Overton and Dhuyvetter, 2020). The number of heifers required and the number generated can be calculated from Equations 1 and 2 (Dechow, 2020).

Equation 1. Number of replacement dairy heifers required:

$$\text{Heifers required} = \text{herd size} \times \text{AFC}/24 \times \text{herd turnover rate} \times (1 - \text{heifer cull rate})$$

where herd size = lactating plus dry cows, AFC = age-at-first calving, herd turnover rate = culling and mortality rate.

Equation 2. Number of replacement dairy heifers generated:

$$\text{Heifers generated} = \text{herd size} \times 12/\text{CI} \times \% \text{ heifers} \times (1 - \text{calf mortality}) \times 24/\text{AFC}$$

where CI = calving interval.

Thus youngstock rearing is an expensive investment, so to reduce costs and optimise genetic potential, good early life management is essential. One of the first steps in cost reduction and improved herd health is having achievable, meaningful performance goals.

3 Targets for heifer rearing

An essential component of youngstock management is data management, data recording, target-setting and benchmarking – if you don't measure, you can't monitor. While ideally the goals of youngstock rearing should be farm-specific, there are generic, within the management system, targets which should be achievable based on the top 25% of farmers' benchmarked performance. Examples of some critical youngstock targets are listed in Table 1. Because of breed variation in some of these standards, relative, rather than absolute, targets can be more useful, for example, achieving double birth weight pre-weaning (8 weeks), 55–60% of mature body weight (MBW) pre-breeding and 85–90% of MBW pre-calving. Targets will vary by country and in some cases can be even more stringent than those listed here, for example, a target of <2% pre-weaning (<3 month) calf mortality per year in the UK (Hyde et al., 2020). Absolute values will also vary by the management system, reflecting their different goals, for example, higher production targets in hyperintensive confinement systems and lower in pasture-based systems (Mee and Boyle, 2020).

While the most appropriate target values are those generated from within the national dairy industry cohort (using data from the top quartile of farmers; norm-referenced thresholds), as opposed to those generated empirically

Table 1 Targets for Holstein-Friesian youngstock rearing

Metric	Target
Perinatal (0–48 h) mortality (%/year)	<5
Calf umbilical infection therapy (%)	<0.5
Calf diarrhoea therapy (%)	<5
Calf respiratory disease therapy (%)	<1
Pre-weaning mortality (%/year)	<3
Heifer mortality (%/year)	<3
Pre-pubertal ADG (kg)	0.6-0.7
Post-pubertal ADG (kg)	0.8-0.9
First breeding age (mo)	13-14
First breeding weight (kg)	325-350
Calving age (mo)	22-24
Calving weight (kg)	550-600
Calving BCS (1-5 scale)	3.25-3.50

(criteria-referenced thresholds), the latter are often more accessible than the former.

In the context of this chapter, setting and achieving targets for youngstock rearing have direct impacts on subsequent dairy herd health as outlined hereunder.

4 Start of the dairy herd health lifecycle

Management of the perinate (and the foetus) is the start of the dairy herd health lifecycle. This management begins prior to calving but has significant effects on young calf health both directly through foetal nutrition and indirectly though effects on colostrogenesis (Abuelo, 2020).

4.1 Effects of pre-calving management of calf health

The data on the effects of pre-calving feeding on neonatal calf health come from both dairy and beef cow studies. Restriction of energy (Carstens et al., 1987) or protein (Funston et al., 2010) during the last third of pregnancy in beef heifers resulted in the birth of calves with reduced thermogenesis and increased susceptibility to hypothermia. Studies in dairy (Nakao et al., 2000) and in beef cows (Bull et al., 1978) attributed reduced perinate vigour ('weak calf syndrome') to pre-calving protein and energy restriction. However, the calves of beef cows in which body condition score (BCS) increased during the dry period and which were in overcondition at calving, took longer to stand and to initiate suckling (Hamilton and Giesen, 1996). Diarrhoea and neonatal calf mortality rates were higher in beef calves born to dams subjected to pre-calving protein and energy restriction (Corah et al., 1975). The diarrhoea may have been due to affected calves being slower to suckle or having reduced immunoglobulin (Ig) absorption. In addition, it has been suggested that diarrhoea in calves from dams with disturbed acid-base balance may be due to altered colostral osmolarity resulting in higher abomasal pH which prolongs casein precipitation time (Gotowiecka et al., 2012). A study in beef cows showed a detrimental effect of maternal protein undernutrition in late pregnancy on calf health after weaning; more treatments for respiratory disease attributed to foetal programming (Larson et al., 2009). Additionally, in recent research, feeding grass silage alone to pregnant dairy cows has been associated with increased incidence of enteritis in dairy calves when compared with feeding grass silage and concentrates (Dunn et al., 2017).

Maternal micronutrient status pre-calving is the major determinant of the micronutrient status and risk of micronutrient disorders in neonatal calves through transplacental (trace elements) and colostral (trace elements and vitamins) transfer of nutrients. Imbalances (deficiency and excess) of micronutrients in calves are associated with micronutrient-specific disorders (e.g. anaemia, CJLD, goitre, nutritional muscular dystrophy). Micronutrient imbalances are also associated with reduced calf immunity and resultant increased incidences of diarrhoea, respiratory diseases and ill thrift, as reviewed by Enjalbert (2009) and Mee (2004). Both the concentration and the chemical form of the micronutrient in the pre-calving diet can influence its effects on the neonate. For example, organic selenium fed to beef cows pre-calving reduced the incidence of diarrhoea in their calves compared to cows fed inorganic selenium (Guyot et al., 2007). While a study in dairy cows showed no detrimental effect of different iodine concentrations in the pre-calving diet on calf health (Connelly et al., 2014), previous studies in sheep had demonstrated negative effects (Boland et al., 2005). Micronutrient disorders are probably of greater significance in suckler beef cows than in dairy cows (Gilmore and Mee, 2012). In addition, a recent study in dairy cows showed that peripartum maternal hypocalcaemia was associated with increased incidence of neonate enteritis compared to calves from eu-calcaemic dams (Hunter, 2015).

4.2 Perinatal management

The critical risks to the calf during the perinatal period (0–48 h) are dystocia, perinatal mortality (from pre-, intra- and post-parturient causes) and infectious diseases (both clinical infections and subclinical infections) (Mee et al., 2013a). Perhaps more importantly, perinatal management has a major impact on subsequent calf health, for example, through the role of colostrum in mitigating the risk of occurrence and the severity of neonatal morbidities (e.g. infectious enteritis and respiratory disease).

Management of each of these critical control points, dystocia, perinatal mortality and infectious diseases is discussed below.

4.2.1 Management of dystocia

Dystocia (defined here as abnormal calving) has both a direct and indirect impact on calf health. It is the primary cause of perinatal mortality. In addition, dystocia indirectly influences the immunity of the newborn calf. This effect is due to the impact of dystocia-induced foetal asphyxiation on reduced colostral Ig absorption and on abnormal perinate behaviour (delayed time from birth to sternal recumbence, standing and sucking). The cumulative effect of these indirect effects of dystocia on the calf is increased risk of failure of passive transfer (FPT) (Furman-Fratczak et al., 2011), increased youngstock morbidity, health therapy, age-at-first calving and mortality (Barrier et al., 2013, Heinrichs et al., 2005). Additionally, dystocia has significant impacts on the dam. Hence, management of dystocia is critical to dairy herd health.

In order to manage dystocia the associated (modifiable) risk factors and causes need to be recognised. The primary causes of dystocia are relative foetal oversize in primiparae and maldispositions (malpostures, malpresentations and malpositions in order of frequency) in pluriparae (Mee, 2008). Given that many of the significant risk factors for dystocia are not under management control (e.g. primiparity, pluriparity, previous dystocia, etc.), farmers need to focus on the modifiable risk factors (Mee et al., 2011).

Management of relative foetal oversize involves ensuring that both foetal body weight and maternal pelvic area are congruent. Thus the modifiable factors associated with calf birth weight such as sire genetic merit for dystocia, breed, nutrition and gestation length need to be managed through genetic selection, feeding management and, where necessary, pharmacological induction of parturition. For Holstein-Friesian cows the optimal calf birth weight range to reduce the risk of dystocia is from 42 kg to 45 kg due to relative foetal oversize (Mee, 2008). Similarly, the modifiable factors associated with the maternal pelvic area such as bodyweight at service and age and body condition at calving, need to be managed through feeding management (Mee, 2014). Neither internal nor external pelvimetry around service is sufficiently predictive of dystocia to select optimal pelvic area-heifers. Of the two approaches, optimising calf birth weight (primarily by genetic selection) is currently considered the more important and feasible long-term management strategy to reduce the occurrence of dystocia in dairy herds. In future, advanced neural network models using recorded cow data (e.g. parity, body condition score, breed, etc.) may be used to predict the likelihood of calving assistance and difficulty; current models have an accuracy of 75% (Fenlon et al., 2017).

Unlike relative foetal oversize, management of dystocia due to maldispositions is more challenging, particularly because the aetiologies of malpostures, malpresentations and malpositions differ. Factors associated with the occurrence of malpostures include pluriparity, prematurity, uterine inertia and foetal ankyloses and mortality. Of these, only uterine inertia secondary to hypocalcaemia can be managed nutritionally to reduce risk of malpostures. Factors associated with the occurrence of malpresentations include older dam age, sire, sire breed, males, lighter foetuses, pluriparity and foetal ankyloses (Holland et al., 1993). Of these, only sire/breed selection can be managed to reduce future risk of occurrence.

Irrespective of the cause of dystocia, good periparturient management can reduce the associated sequelae. Key management practices are prediction of calving time, good calving supervision and good obstetrical technique. The impacts of night-time feeding of pregnant cows to reduce dystocia associated with night calvings are equivocal (Gleeson et al., 2007). Ongoing formal training of farm staff periparturient management can improve outcomes.

4.2.2 Management of perinatal mortality

The prevalence of perinatal mortality can vary internationally between 2 and almost 10%. Given the newborn calf has the best genetics in the herd, the loss of such valuable genetic material for future productivity needs to be minimised. In addressing the issue of perinatal mortality one needs to disaggregate the losses into those which occur prepartum, peripartum and postpartum, as different risk factors, aetiologies and management strategies apply in each category of loss. The primary causes of perinatal mortality are non-infectious and vary by calving assistance score with anoxia and maldispositions the most common causes of death. An exemplar dataset from Irish commercial dairy herds is shown in Table 2 where the predominant cow breed is Holstein-Friesian but there are also, in decreasing order, Jersey and Norwegian Reds, and their cross breeds, and the predominant sire breed is Holstein-Friesian but there are also, in decreasing order, Jersey, Aberdeen Angus, Norwegian Red, Hereford and other beef breed sires (Belgian Blue, Simmental, Limousin, etc.).

Thus management to reduce perinatal mortality should focus on the dystocia-associated modifiable risk factors for stillbirth/perinatal mortality (Mee et al., 2014) and asphyxia during calving. However, given the variability between farms, and years in the causes of mortality, investigation of losses (Mock et al., 2020) should be a standard protocol on all dairy farms.

Foetal asphyxiation is a major cause of perinatal mortality; hence, management of this problem is recommended. Approaches include shortening the duration of stage two of calving (by avoiding disturbance or movement of animals during late stage one of calving, judiciously timed intervention during stage two or promptly performing a Caesarean section) and calf resuscitation (Mee, 2018).

Other causes of perinatal mortality can be managed as appropriate to the cause, for example, prepartum macronutrient and micronutrient supplementation of the dam (Mee, 2014). However, in herds with high perinatal mortality, farmers and their veterinarians need to focus on periparturient management (Mee et al., 2013b).

4.2.3 Management of perinatal infections

Perinatal infections need to be categorised into those which occur *in utero* and after birth as their management differs.

4.2.3.1 In utero infections

In utero infections occur in approximately 20% of perinatal mortalities and contribute to death in approximately 10% of cases (Jawor et al., 2017). Common foetopathogens include *Neospora caninum*, bovine viral diarrhoea

Table 2 Causes of perinatal mortality (0–48 h) (%) categorised by calving assistance score (n=259) (Mee and Kenneally, 2017)

Cause of calf death (%)	Unobserved calving (n=70)	Observed, no assistance (n=46)	Easy assistance (n=57)	Moderate difficulty (n=47)	Severe difficulty (n=39)	All calvings (n=259)
Anoxia	30	17.4	14	12.8	20.5	19.7
Mal-disposition	0	2.2	17.5	34	43.6	17
Congenital defect(s)	10	21.7	12.3	10.6	12.8	13.1
Haemorrhage/Anaemia	14.3	15.2	10.5	2.1	2.6	9.7
Premature placental separation	1.4	2.2	19.3	23.4	2.6	9.7
Other	22.9	21.7	14.1	12.8	17.9	18.1
No significant findings	21.4	19.6	12.3	4.3	0	12.7

(BVD) virus, *Leptospira*, *Listeria*, *Escherichia*, *Trueperella* and *Salmonellae* spp. and pathogenic fungi. While for some of these infections management is possible, for others (e.g. *Escherichia*, *Trueperalla*, mycoses) the organisms are ubiquitous or so sporadic that effective management is not possible. Control of neosporosis involves necropsy examination/sampling (PCR) of all foetal and perinatal calf mortalities, herd screening (e.g. bulk tank milk serology) to establish likely prevalence, whole herd serology to detect positive animals and culling or breeding to beef sires of detected cows. In addition, implementation of both effective bioexclusion (e.g. blood testing purchased animal), and biocontainment measures (e.g. preventing canid faeces from getting into the cows' feed) are recommended (Mee et al., 2012). If a national BVD eradication programme is not in place, sampling of all newborn calves (e.g. ear tag biopsy), herd screening, bioexclusion and vaccination are recommended to manage this disease. For pathogenic *Leptospira* and *Salmonella* spp., vaccination is recommended both to control disease in calves but also to reduce associated zoonotic risks.

4.2.3.2 Neonatal infections

Infections contracted by newborn calves include both those likely to result in clinical signs of enteritis, septicaemia (e.g. *Escherichia coli*, *Salmonella* spp.), omphalophlebitis, arthritis, meningitis and subclinical infections (e.g. *Mycobacterium bovis* subspecies *paratuberculosis* (MAP) and *Mycoplasma bovis*). These neonatal infections can have long-term consequences for youngstock growth, productivity and health (vide infra).

Central to control of all neonate infections is reducing infectious challenge, for example, environmental hygiene in the calving and calf housing and calf feeding equipment and improving host immunity, for example, passive transfer of immunity via best practice colostrum management.

4.2.4 Colostrum management

Providing an adequate volume of hygienic, good quality, first milking colostrum as soon as possible after calving to the neonatal calf is the bedrock of good colostrum management. Numerous mnemonics have been devised to embed this axiom in farmer's management standard operating procedures (SOPs), for example, *Colostrum 1-2-3*; colostrum is only the first (1) milking which should be given within two (2) hours of birth in a feed of three (3) or more litres (body weight-dependent) (Animal Health Ireland). In the absence of any/sufficient natural colostrum, colostrum replacers or supplements have a role to play especially if the cow's colostrum quality or quantity is inadequate, or in Johne's test-positive herds, and, critically, the product, and number of doses, provides the necessary amount of total Ig (Williams et al., 2014, Mee et al.,

1996). While the roles of colostrum in preventing calf diseases and improving calf growth are well established, recently the role of colostrum in transferring signalling molecules (microRNAs) which are important for neonatal intestinal development and the complete immune system has been identified (Van Hese et al., 2020). Harvesting of good-quality colostrum begins with the management of the pregnant cow.

4.2.4.1 Effects of pre-calving management on colostrogenesis

The studies on the effects of pre-calving feeding of cows on colostrum quantity and quality are derived from dairy and beef cow studies. Altering dietary energy content or restricting DMI during the far-off (Nowak et al., 2012a) or close-up dry period (Nowak et al., 2012b) in dairy cows had no effect on colostrum or calf serum Ig concentrations. Feeding diets high in undegradable protein during the dry period to dairy cows did not alter colostrum or calf serum Ig concentrations (Bland et al., 2007). However, a negative correlation was found between the amount of concentrates fed to dairy cows pre-calving and colostrum IgG content by Gulliksen et al. (2008) attributed to a possible dilution effect with increased colostrum yield. Earlier studies by Stockdale and Smith (2004) in dairy and Shell et al. (1995) in beef cows found a similar effect with increasing energy intake/plane of nutrition during gestation. Stockdale and Smith (2004) also found that increasing dietary crude protein content increased colostrum Ig content in dairy cows. Both Santos et al. (2001) and Hook et al. (1989) found no effect of diet crude protein content and colostrum volume or IgG content in dairy cows. Similar results on diet crude protein content were found in beef cattle (Quigley and Drewry, 1998, Funston et al., 2010) though the latter authors did find lower serum IgG content in calves from dietary protein-restricted beef cows. While beef cows fed straw only in the last fortnight pre-calving had lower colostrum and calf serum IgG content (McGee et al., 2006), in general dietary nutrient restriction or supplementation did not alter colostrum quantity and quality (Quigley and Drewry, 1998, Hough et al., 1990, Burton et al., 1984). However, one recent study in dairy cattle found increased colostrum yield (though not Ig concentration) in cows fed grass silage and concentrates pre-calving compared to those fed grass silage alone (Dunn et al., 2017). Ig absorption was reduced in calves fed colostrum from restricted dairy (Burton et al., 1984) and beef cows (Rytkonen et al., 2004, Hough et al., 1990). This may be associated with lower colostral tri-iodothyronine, which is involved in Ig absorption, or altered foetal intestinal vascularity and foetal enterocyte proliferation, in restricted cows. A low-energy diet fed pre-calving, though reducing colostrum volume, increased colostral Ig concentration (Mann et al., 2013). Interestingly, 'focus feeding' in sheep has been shown to increase colostrum volume (Banchero et al., 2004), though effects on Ig content were not reported; similar results are not available in dairy or beef cows. Selenium supplementation of the pre-calving diet has been shown to increase colostrum volume and Ig content (Mee, 2004) and calf serum Ig content (Illek, 2010). For example, selenium supplementation of pregnant cows increased calf Ig absorption efficiency and hence calf serum Ig content (Hall et al., 2014). However, parenteral selenium administration to dairy cows (Leyan et al., 2004) and high dietary iodine (Boland et al., 2005) and selenium content (Hammer et al., 2007) in pregnant ewes has been shown to reduce offspring serum Ig content. Supplementing the diet of pre-calving cows with mannan oligosaccharides tended to improve calf colostrum-derived rotavirus antibody titres in response to maternal vaccination (Franklin et al., 2005). There are conflicting results on the effects of feeding anionic diets to dairy cows pre-calving on calf acidosis, colostrum Ig absorption and serum Ig concentration (Morrill et al., 2010, Quigley and Drewry, 1998).

In addition to these nutritional management factors, farmers can increase the quality of the colostrum they feed to their calves in other ways. Vaccination of the dam pre-calving against common calfhood enteropathogens (e.g. rotavirus) can produce pathogen-specific, hyperimmune colostrum (Civra et al., 2019). Other factors such as dam parity and temperature-humidity-index (THI) can also affect colostral Ig content (Shivley et al., 2018). But perhaps of more practical import is the fact that short dry periods (less than 8 weeks) reduce colostrum quality and colostrum Ig content decreases by approximately 4% for every hour after calving that the cow is not milked (Morin et al., 2010). These factors emphasise the importance of colostrum-focussed dry cow and periparturient management in enhancing colostrum quality for the newborn calf. Unfortunately the measurement of colostrum quality is not common on dairies. While the 'gold standard' colostrum immunity quality test is the measurement of the colostrum immunoglobulin G concentration by single radial immunodiffusion (SRID), this is an expensive laboratory assay only used in research studies. However, surrogate tests have been validated against this assay which can be used by farmers (Quigley et al., 2013). One of the most commonly used tests is refractometry. This can be carried out with a Brix optical refractometer. This inexpensive device has the advantage over a colostrometer (which measures colostrum specific gravity as a proxy for total protein content) that colostrum temperature or foaming (when a foam top develops on the colostrum sample after pouring) does not affect the reading. Good-quality colostrum has a Brix value of greater than or equal to 21%, equivalent to an SRID value of 50 g/L of IgG (Quigley et al., 2013). A recent review indicated that only 10–20% of farmers internationally implement this simple practice of measuring colostrum quality (Mee, 2020).

4.2.4.2 Colostrum hygiene

While producing an adequate volume of high Ig content colostrum is a basic management goal, this good work can be undone by bad colostrum hygiene. Microbial contamination of colostrum can impair Ig absorption, accelerate gut closure and cause diarrhoea or sepsis and reduced weight gain in neonatal calves. Such contamination is surprisingly common. For example, recent surveys of the colostrum microbiome showed that only a quarter of samples met the total plate count target while over 75% showed evidence of faecal contamination (Nejedla et al., 2019, McAloon et al., 2016). Thus, management of neonatal calf immunity begins with a collection of high microbial quality colostrum and either refrigeration for short term (a day) or freezing for long-term (up to a year) storage or heat treatment (pasteurisation) to ensure 'clean' colostrum is fed to calves.

4.2.4.3 Failure of passive transfer (FPT)

Newborn calves do not have an active immune system and so, colostral antibodies are the only source of immunoglobulins to protect them from infectious diseases. Method of colostrum feeding (e.g. suckling vs. bucket feeding vs. stomach tubing) can affect the risk of FPT (suckling>stomach tubing>nipple bucket feeding; but, volume-dependent) (Godden et al., 2009). However, one study found no effect on subsequent clinical health scores or average daily gain (ADG), (Ebert et al., 2007).

Measurement of calf blood Ig concentrations (or surrogate measures, for example, total protein TP, ZST, Brix readings) (Hogan et al., 2015) can be used as a management tool to screen (on a group, not an individual basis) for adequacy of passive transfer (in healthy calves 2-7 days old) and to predict future calf health and growth (Cuttance et al., 2018). For example, calves with a TP <50 g/L had a 2.4-times greater chance of dying than calves with a concentration between 50 and 60 g/L (McCorquodale et al., 2013). Recommended targets include 70% of samples above 60 g/L for TP and above 20 units for ZST. A recent study concluded that to completely avoid FPT, calves should receive at least 2.5 L of high-quality colostrum (Ig >88 g/L) within one hour of birth (Lora et al., 2018). Though not the focus of this chapter, male calves, whose health and welfare should be treated similarly to that of female calves, are at greater risk of FPT through receiving a lower volume of colostrum and of receiving pooled colostrum (Renaud et al., 2020).

4.2.4.4 Snatch calving

In addition to avoiding pooled colostrum (colostrum from more than one cow), implementation of 'snatch calving' (immediate removal of the calf from the cow after calving) is currently recommended to reduce the risk of transmission of MAP between the cow and her calf. However, this practice is coming under scrutiny due to the associated welfare implications for the cow-calf dyad (Beaver et al., 2019).

It is clear from this overview that ensuring that calves survive the perinatal period and do not suffer detrimental sequelae due to suboptimal neonatal management, especially FPT, can best prepare the calf for subsequent development, growth and health. However, both calfhood nutritional management and diseases can undermine the good work carried out in the perinatal period.

5 Impacts of calfhood nutritional management on subsequent dairy herd health

There is now an extensive corpus of literature showing that management of the neonatal and later developmental stages of the calf's life impacts subsequent dairy herd health, age-at-first calving, lifetime productivity and survival. For the latest review, see Hammon et al. (2020). The latest adjacent literature on the effects of management of the pregnant cow (e.g. milk production level, heat stress, ration energy density) on foetal developmental programming ('The Barker Hypothesis'; Barker et al., 1989), specifically, immune function (Cooke, 2019) and youngstock health (Perry et al., 2019), support this hypothesis. Nutritional programming of calves alters their long-term metabolic and productive performance as adults. This is a permanent effect on life-long gene expression of the release of hypothalamic neuropeptides which control feed intake and weight gain (Taylor and Poston, 2007). These related streams of research, foetal and calfhood programming, have altered the paradigm of management as a factor just affecting current performance to one potentially affecting future performance also. Specifically in relation to milk feeding, this research has highlighted the limitations of restricted milk supply and demonstrated the short- and long-term benefits of intensive/accelerated milk feeding protocols (Hammon et al., 2020, Hu et al., 2020). However, despite most dairy farmers rearing their own replacement heifers, the impacts of calfhood management and disease on subsequent productivity and longevity are often unrecognised (or unknown, for example, the lactocrine hypothesis; Bagnella and Bartol, 2019) by both farmers and their veterinarians. This has implications for farmer perception and management of calfhood nutrition and health hazards and veterinary communication on calfhood nutritional and disease sequelae with lag dynamics.

5.1 Age at first calving (AFC)

Age at first calving (AFC) is important because it affects farm profitability through heifer rearing costs and effects on subsequent reproductive and productive performance. AFC is correlated with milk production (Eastham et al., 2018), calving (Mee, 2008) and reproductive performance (Ettema and Santos, 2004) and longevity (Berry and Cromie, 2009). There is a linear association between AFC and first lactation peak and 305-day milk yield (Berry and Cromie, 2009). Target/optimal values for AFC vary widely internationally between 21 (Cole et al., 2013) and 30 months (Haworth et al., 2008), however, a recent knowledge summary literature review concluded that the optimum range should be narrower; from 22 months to 25 months (Steele, 2020). Actual median values vary between 23 months (Stanton et al., 2012) and 29 months (Haworth et al., 2008). Lower AFC is associated with better udder health, better reproductive performance and increased lifetime daily milk production (Eastham et al., 2018). Given the time lag between calfhood nutritional management and first calving, farmers or their veterinary practitioners are rarely aware of the potential impacts of former on the latter.

5.1.1 Calf nutritional management factors affecting age at first calving (AFC)

Feeding a higher volume of colostrum has been shown to be associated with a lower AFC in some (Mohd Nor et al., 2013) but not in other studies (Faber et al., 2005; DeNise et al., 1989). However, feeding a higher volume of colostrum (Brickell et al., 2009) and higher serum Ig concentrations after birth (Robison et al., 1988) have been shown to significantly increase the ADG and girth in the first six months of life, factors which reduce AFC. For example, a recent study showed an inverse linear relationship between ADG of heifers at the start of the breeding season and subsequent days open (Hayes et al., 2019). Additionally, calves with higher serum Ig concentrations at 1-2 days of age had a lower age at insemination (Furman-Fratczak et al., 2011), which affects AFC. Not feeding waste milk to calves has been associated with increasing the odds (OR 3.656; CI95 0.953-14.034, $P=0.059$) of heifers calving at $\leq$24 months compared to >24-<27 months (Mohd Nor et al., 2013).

Higher nutrient intake pre-weaning has been shown to affect AFC. Increasing the volume of milk fed increased the odds (OR 1.776; CI95 0.932-3.384, $P=0.081$) of heifers calving at $\leq$24 months compared to >24-<27 months (Mohd Nor et al., 2013). This is in agreement with previous data from the United States where maximum milk intake positively influenced AFC (Heinrichs et al., 2005) and from the UK where ad libitum milk feeding significantly increased ADG and heifer height from 1 to 6 months of age compared to restricted feeding (Brickell et al., 2009). However, a high ADG (1.15 kg/d) prior to breeding (8-10 months) delayed the onset of puberty (Silper et al., 2013) which could increase the AFC. In addition, heifers fed poorer quality forage (higher ADF) tended to have increased AFC ($P=0.081$)

(Heinrichs et al., 2005) and heifers in Sweden which spent more time grazing had increased AFC (Hultgren et al., 2008).

Only a limited number of studies compared the effect of whole milk (WM) versus milk replacer (MR) on AFC. In a Dutch study the median AFC was lower for calves fed MR compared to WM (P=0.012) (Mohd Nor et al., 2013). This is in agreement with a study from the UK where calves fed MR had significantly increased ADG, girth, crown-rump length and height from 1 to 6 months of age compared to calves fed WM (Brickell et al., 2009). In contrast, in Israeli studies age at puberty was lower (P<0.01) (Shamay et al., 2005) and AFC tended to be lower where calves were fed WM compared to MR (P<0.07) (Moallem et al., 2010). Differences in the nutrient composition and feeding regimes of MR between studies probably account for these differences.

The effects of milk replacer feeding regimes on AFC are also conflicting. While increasing MR feeding level (g/day and crude protein (CP) content) tended to reduce both age at first observed oestrus and age at first service, AFC was not altered in a Northern Ireland study (Morrison et al., 2012). Similar results were found by Terre et al. (2009) in a Spanish study where feeding a higher volume of a high dry matter (DM) MR increased BW at weaning but not thereafter or age at first breeding, compared to a lower feeding rate of a lower DM MR. Drackley et al. (2007) did not change AFC by utilising a conventional limit-fed or an intensified MR programme. In contrast, a high-CP, high-feeding volume MR regime reduced AFC numerically (P=0.11) compared to a conventional MR regime (Davis Rincker et al., 2011) and an intensive high-solid (16.7%), high-feeding MR regime reduced AFC compared to a conventional MR feeding regime (P=0.05) (Raeth-Knight et al., 2009). It may be concluded that reduced AFC is associated with increased nutrient intake and ADG pre-weaning.

5.2 Lifetime productivity

The lifetime productivity (defined as from first calving to death) of a cow is short (3-4 years) relative to natural life expectancy (~20 years); this is a welfare concern (De Vries and Marcondes, 2020). It is influenced by many interrelated factors such as age at first calving, calving intervals, milk yield/day, availability of replacement heifers, and age at removal but also by nutritional management of the pre-weaned calf. Heifers that calve at a younger age have more days of productive life (higher longevity index) and higher lifetime milk yields (Lin et al., 1988). A heifer that first calves at three years of age and is culled in her third lactation spends only 40% of her life being productive, whereas one that calves at two years and survives to her fifth lactation produces milk for more than 60% of her life thus increasing her lifetime productivity (Garnsworthy, 2012). Given the high (30%, Mohd Nor et al., 2012) and increasing (Bach, 2011) culling, and hence, replacement rates on modern dairy farms it is becoming increasingly difficult for farmers to increase cow lifetime productivity profitably. Hence any impacts of calfhood management on lifetime productivity are increasingly important.

5.2.1 Calf nutritional management factors affecting lifetime productivity

Given the multitude of nutritional, genetic and environmental factors which impact lifetime productivity it can be difficult to detect early life effects such as dairy calf nutritional management. However, such influences have been reported. The beneficial effects of feeding a larger volume of colostrum (4 vs. 2 L) (Faber et al., 2005) and of higher serum Ig status (DeNise et al., 1989) have been shown for first and second lactation milk production. Higher nutrient intake pre-weaning has been reported to affect lifetime productivity though results are conflicting. Drackley et al. (2007) found a response in first lactation milk production to an intensive MR programme ($P<0.01$) and an Israeli study found that calves offered ad libitum WM compared to MR produced 1.3 kg/d more 3.5% fat-corrected milk in first lactation ($P<0.005$) (Shamay et al., 2005). However, Morrison et al. (2012), Raeth-Knight et al. (2009) and Terre et al. (2009) failed to find significant effects of enhanced milk feeding regimes. Gleeson and O'Brien (2012) found no effect of milk feeding frequency (once vs. twice daily) on ADG or first lactation milk production. A review stated that greater nutrient supply through increased volume of milk appears to increase milk yield, but more research is needed to confirm this effect (Khan et al., 2011). This opinion was supported by a meta-analysis which concluded that calves fed high nutrient liquid diets were twice more likely to have greater milk yield in first lactation (Soberon and Van Amburgh, 2013).

The ADG between birth and breeding can affect lifetime productivity. Soberon et al. (2012) showed that pre-weaning ADG, weaning weight ($P<0.03$) and ADG from birth to breeding ($P<0.01$) was positively correlated with first lactation milk yield. For every kg of pre-weaning ADG, heifers produced 850 kg more milk during their first lactation. Significant positive correlations were also found for ADG and milk production in lactations one through three indicating impacts on lifetime productivity. The authors concluded that metabolic programming during early life has lifelong implications for milk production possibly through epigenetic changes. The nutritional management of the calf pre-weaning has the greatest impact on the development of the mammary gland which is associated with higher subsequent milk production (Geiger et al., 2016). It is also possible that a lactocrine effect of bioactive factors in WM affects milk yield as Moallem et al. (2010) reported that calves fed WM had a 10% higher first lactation milk yield than those fed MR ($P<0.001$). These findings are supported by those of Heinrichs and Heinrichs (2011) who found that for every 1 kg increase in DMI at weaning, there was a corresponding increase of 287 kg in first lactation milk yield. While decreased mammary development has been observed as prepubertal ADG is increased, a meta-analysis by Zanton and Heinrichs (2005) concluded that first lactation milk production is increased as prepubertal ADG (150–320 kg) increased up to 799 g/d whereas further increases in ADG were associated with lower milk yield indicating a curvilinear relationship. A meta-analysis concluded that for every kg of pre-weaning ADG, first lactation milk yield increased by 1550 kg (Soberon and Van Amburgh, 2013). The beneficial effects of larger, well-grown heifers on milk production were also shown by Archbold et al. (2012) who reported that heavier heifers and those in better BCS at mating start date in a seasonal calving system had significantly higher milk production in lactation one through three indicating effects on lifetime productivity.

6 Impacts of calfhood diseases on subsequent dairy herd health

Similar to the effects of youngstock nutritional management, calfhood disease management has both short- and long-term impacts on subsequent dairy herd health and performance. Calfhood disease can reduce the benefits of nutritional programming, with calves which develop respiratory diseases having poor subsequent average daily gain despite earlier intensive feeding management (Maccari et al., 2015).

6.1 Age at first calving (AFC)

Three aspects of youngstock health management have been shown to affect AFC, calving management, calf diarrhoea management and respiratory disease management.

Dystocia has short-term detrimental effects on calf health such as increasing the risk of FPT or incidence of umbilical cord pathologies (Moscuzza et al., 2014). However, often not recognised are the long-term sequelae. For example, a North American study showed that as calving assistance score of a calf increased by one unit, her subsequent AFC increased by 17 days ($P=0.048$) (Heinrichs et al., 2005).

The effects of calf diarrhoea on AFC are conflicting. Some studies (e.g. Morrison et al., 2010; Hultgren et al., 2008) found no influence of diarrhoea pre-weaning on AFC. In contrast, there is a tendency ($P=0.064$) for calves treated for diarrhoea or pneumonia to have increased AFC in a study by Heinrichs et al. (2005), and Waltner-Toews et al. (1986) found that calves treated for diarrhoea pre-weaning had a higher AFC.

The effects of bovine respiratory disease (BRD) on AFC are generally reported as detrimental. For example, Stanton et al. (2012) in Canada showed that calves with BRD were 0.8-times as likely to calve as heifers without BRD (P=0.02) and the odds of calving by 25 months of age were 0.6-times lower in calves with BRD (P=0.01). Donovan et al. (1998) had shown that days treated for BRD before 6 months of age decreased heifer ADG between birth and 6 months of age and between 6 and 14 months of age (P<0.025). This is in agreement with earlier studies which showed that calves with BRD were less likely to calve and were calved later (Warnick et al., 1994; Correa et al., 1988; Waltner-Toews et al., 1986). In contrast, Morrison et al. (2010) found no effect of BRD pre- or post-weaning on AFC in one study in Northern Ireland. Interestingly, high-housing maximum relative humidity (P=0.032) and mean ambient temperature (P=0.028) were both associated with increased AFC in one North American study (Heinrichs et al., 2005).

6.2 Lifetime productivity

Similar to effects on AFC, calving management and calfhood morbidity have been shown to detrimentally impact dairy herd productivity.

The more difficult a heifer calf's birth, the lower her subsequent milk production. As calving assistance score of a calf increased by one unit, her first lactation milk production decreased by 285 kg (P=0.03) and lifetime production tended to decrease (P=0.06) in one North American study (Heinrichs and Heinrichs, 2011). Previous studies had shown that the risk of FPT was significantly higher (OR = 2.6) for calves on operations where farmstaff would not seek veterinary assistance when unable to correct a calf during dystocia (Beam et al., 2009). In addition, a recent study has shown that calves which survive dystocia experience greater physiological stress, lower passive immunity transfer and higher mortality (Barrier et al., 2013). Dystocia-associated FPT may increase the risk of digestive and respiratory diseases in calves (Lombard et al., 2007), which may in turn contribute to lower subsequent productivity.

Most studies did not find a significant influence of calf diarrhoea on milk production (Soberon et al., 2012; Morrison et al., 2010; Warnick et al., 1995; Britney et al., 1984). However, one Swedish study found that calves that had diarrhoea during their first 3 months of life had 344 kg lower milk production in first lactation (Svensson and Hultgren, 2008).

The evidence for effects of calfhood BRD on milk yield is more convincing. Morrison et al. (2010) found that BRD pre-weaning tended (P=0.10) to and BRD post-weaning did (P<0.05) decrease first lactation milk production (by 6%). The same group reported that while calves treated for a single episode of BRD did not have lower milk yield, calves treated for multiple cases of BRD had lower first (P<0.05) and second lactation (P<0.01) milk production (Morrison et al., 2013). A Canadian study showed that first test milk production tended to be 1.1 kg lower for heifers with BRD (P=0.06) but 305-day milk production was not affected (Stanton et al., 2012). A Swedish study reported an association between severe BRD before 90 days of age and lower first lactation milk production ($P=0.02$) (Svensson and Hultgren, 2006). Perhaps, not surprisingly, vaccinating against bovine respiratory syncytial virus (BRSV) was shown in one study to be associated with an increase in first lactation milk production of 493 kg ($P=0.063$) (Mohd Nor et al., 2013). Earlier studies by Warnick et al. (1995) and Britney et al. (1984) did not show an association between BRD and milk production.

However, a number of studies have shown that calf morbidity is associated with lower milk production. Soberon et al. (2012) showed that calves receiving antibiotics (primarily for BRD) had 493 kg lower milk yield in first lactation ($P<0.01$). Similarly, Heinrichs and Heinrichs (2011) reported that the number of days that calves had diarrhoea or BRD in the first 4 months of life had a negative effect on first lactation milk production ($P<0.01$). 'Other' calf health events were associated with lower milk yield in first lactation in an earlier study by Miller and Faust (2000).

6.3 Survival

Calving management and calf-hood morbidity, especially BRD, have been shown to significantly reduce heifer survival in the dairy herd.

Calves which had a difficult calving had a three-fold greater risk of dying before weaning ($P<0.05$) and before first service ($P<0.001$) compared with heifers born unassisted. Survival to first calving was not affected by the difficulty experienced at birth (Barrier et al., 2012). Calves with poor neonatal immunity are less likely to survive. McCorquodale et al. (2013) found that calves with serum total protein (TP) values of <50 g/L were 2.4-times more likely to die within 4 months than calves with higher TP concentrations. These results are in agreement with earlier findings on survival up to six months of age (Robison et al., 1988).

Heifer calves which developed omphalo-arthritis had significantly poorer survival distribution up to the second lactation than unaffected calves (Britney et al., 1984). Disease treatment during the first week of life was associated with a 1.3-times greater risk of mortality and calves that received three or more treatments were 3.5-times more likely to die before 4 months of age (McCorquodale et al., 2013). However, the number of days of illness did not affect survival or culling age (Heinrichs and Heinrichs, 2011) and calfhood disease was not associated with length of herd life in an earlier American study (Warnick et al., 1997).

The most consistent relationships between calf health and survival were reported for BRD (Van Der Fels-Klerx et al., 2002). In a Canadian study only 66% of calves with BRD survived to first calving compared to 84% without BRD.

However, BRD did not affect survival to 120 DIM for calved heifers (Stanton et al., 2012). Similar detrimental effects of BRD on survival to first calving were reported by Warnick et al. (1995) and by Correa et al. (1988). These findings are consistent with those of Waltner-Toews et al. (1986) who found that heifers treated for BRD in the first 3 months of life were 2.5-times more likely to die between 3 months of age and calving. While Bach (2011) found no association between incidence of BRD and survival they did find that heifers that experienced four or more cases of BRD had greater odds (OR 1.87) of not surviving the first lactation. In addition, the accumulated DIM throughout productive life and the proportion of productive days with respect to recorded days of life decreased linearly ($P<0.05$) as the number of BRD episodes increased. Previously Moriarty et al. (2007) had shown the association between calfhood BRD and necropsy lesions of unresolved pneumonia and stunting due to lack of compensatory growth in later life.

Given these relationships between calfhood health traits and future dairy herd health it would appear logical to incorporate calfhood traits in future genetic selection indices, as has been suggested (Mousa et al., 2015). However, a recent study indicated that different SNP markers contributed to disease in calves and in cows; hence, such selection may not be appropriate with current knowledge (Mahmoud et al., 2017).

7 Role of vet in communicating best practice in youngstock management

Veterinary practitioners play a central role in better youngstock management ideally as part of a herd health and productivity management programme (Boersma et al., 2010). Farmers expect their consultants to identify calf care issues, inform farmers about the issue and provide practical steps to remediate these issues (Croyle et al., 2019). One of the issues veterinary practitioners and consultants need to be cognisant of is farm-blindness. Farm-blindness may be defined as 'a misperception by farmers that what they see every day on their own farm is normal, particularly when it is not; a new normal' (Mee, 2020). This paradigm applies to all aspects of farm management including other important animal health issues, for example, lameness. Evidence of farm-blindness in youngstock management may be found in failure to implement recommended practices, for example, prompt colostrum feeding, or failure to recognise high calf morbidity or mortality rates. Farmers' youngstock management hazard ranking perceptions differ significantly from those of their veterinarians; hence, the latter need to recognise this asymmetry when addressing the core issue of farm blindness. De-normalising poor youngstock management can be achieved by first creating awareness of the existence of the problem (e.g. through external audit of farm performance and national calf health campaigns. e.g. CalfCare[R] – Animal Health Ireland), secondly, by providing farmers with locally relevant peer benchmarks against which to compare their own youngstock performance (e.g. re-norming using regional/national calf mortality rates), thirdly, by using agri-technology to provide farmer-independent alerts to deviations in animal performance and finally, effectively communicating current recommended best practices to farmers (e.g. good colostrum management) through peers, practitioners or domain experts.

A critical concept veterinarians need to grasp is the principle that with individual animal cases the treatment protocol is under the control of the vet. However, with herd problems, the resolution of the issue resides with the farmer (More et al., 2017). Hence, effective communication of prioritised, best-practice recommendations (in particular what good farmers are doing – use of social contagion herd effect nudges) is critical to enable most farmers to realign their management with current norms. Veterinary trouble-shooting of problems in animals and herds will still be required but the central role of the farmer needs to be recognised both by the vet and by the farmer. Realignment needs to address both activity (processes) and performance (output). Veterinarians and specialised advisory staff have a critical role in extending calf-rearing messages to farmers (Svensson and Hultgren, 2008) but must be cognisant of the limitations of one-step thinking which may miss the more subtle web of causality. Thinking in knowledge transfer is changing significantly from the former linear, top-down approach to a more sharing, bottom-up, design solutions approach (Hennessy and Heanue, 2012). Peer-to-peer learning is particularly effective in this regard via farmer action or discussion groups (Morgans et al., 2018; Hennessy and Heanue, 2012). As farms get larger and evolve from an owner-operator model to a distributed model, staff numbers/job titles increase. Inter-staff communication becomes more complex as the goals of care staff may be process-orientated (e.g. calf hutch hygiene) while those of managers are outcome-orientated (e.g. calf mortality rate) (Pereira et al., 2014). Also, educational, generational and language barriers may need to be overcome. Effective communication results in better stockmanship and good calf management stockmanship knowledge, attitudes and perceptions (KAPs) are positively associated with better calf health outcomes (Adler et al., 2019; Vaarst and Sorensen, 2009). Novel models of farming may also contribute to better youngstock management where calves are sent to specialised/custom contract rearers until the point of calving (Mee et al., 2018). In addition to addressing issues of poor youngstock management, education and public engagement (EPE) on best practice in youngstock management will be a pillar of sustainable agricultural systems ('social license to farm') in the future to avoid reputational risk to the dairy industry.

8 Conclusion and future trends

The material reviewed in this chapter clearly demonstrates that youngstock management can play a critical role in optimising dairy herd health. While the immediate impact of better calf management is visible to farmers in better youngstock health and growth, the long-term benefits in dairy herd productivity and survival need to be emphasised more by veterinarians and agricultural advisers.

The most consistently reported management factors significantly associated with reduced age at first calving (AFC), higher milk production and longer survival were feeding a larger volume of colostrum, offering a higher liquid feed volume pre-weaning, offering a higher solids milk replacer, offering whole milk and a higher average daily gain. The three health disorders most frequently associated with increased AFC, lower milk production and shorter survival were dystocia, calf diarrhoea and respiratory disease. It is concluded from this review that dairy calf health and heifer management can significantly affect age at first calving, lifetime productivity and survival. These results have implications for producer perception and management of calfhood nutrition and health hazards and veterinary communication on calfhood disease sequelae with lag dynamics.

Based on recent changes in the dairy industry worldwide, it is predicted that some of the important future trends in youngstock management will include more emphasis on reducing antimicrobial use in calf rearing (and perhaps increased emphasis on direct-fed microbials and prebiotics and probiotics, for example, as colostrum, supplements), improving calf welfare by external certification ('happy calves'; calf comfort will be the new cow comfort), incorporation of precision livestock farming concepts in calf rearing for pre-clinical diagnostics (e.g. wearable wellness sensors), increased use of SOPs for calf/heifer management on 'traditional' smaller dairy farms, the growth of specialised vertically integrated contract calf/heifers rearing and increased use of social science to 'nudge' stakeholders towards best practice of youngstock management. National infectious disease eradication schemes (e.g. bovine viral diarrhoea virus and possibly *Mycoplasma bovis*) will tangentially improve calf immunity and health, while if gene editing is approved in food-producing animals the resultant chimeras will dramatically change calf health.

9 Where to look for further information

9.1 Further reading

- Farm health and productivity management of dairy young stock by S.-J. Boersema, J. Cannas da Silva, J. F. Mee and J. Noordhuizen.

- Dairy heifer management, Veterinary Clinics of North America, Edited by S. Godden and S. M. McGuirk.

9.2 Key journals/conferences

- Journal of Dairy Science – international peer-reviewed research results/ reviews; researcher-focused.
- World Buiatrics Congress – international cattle disease congress; veterinary practitioner-focused.
- Smart Calf Rearing Conference – international calf rearing/health conference; farmer/industry-focused.
- Healthy Calf Conference – Canadian veal calf rearing and health conference; farmer/industry-focused.
- Dairy Calf & Heifer Association Conference – North American calf rearing and health conference; farmer/industry-focused.
- European Calf Conference (ECC) – European calf rearing and health conference; farmer/industry-focused.

9.3 Youngstock associations internationally

- Calfcare – Irish national calf rearing and health technical working group (www.animalhealthireland.ie) – stakeholder-focused.
- KalfOK – Dutch national calf management programme (www.gddiergezondheid.nl) – farmer-focused.
- DCHA – American Dairy Calf & Heifer Association – national calf rearing and health organisation (www.calfandheifer.org) – farmer-focused.
- Healthy Calves – New Zealand commercial calf health advice (www.healthycalves.co.nz) – farmer-focused.
- CalfNotes – American calf nutrition and rearing advice (www.calfnotes.com) – farmer-focused.
- Calf Care – Canadian calf rearing organisation (www.calfcare.ca) – farmer-focused.
- CalfMatters – UK commercial calf health plan advice (www.calfmatters.co.uk) – farmer-focused.

9.4 Major international research projects

- The BRD 10K study – a major North American research study of respiratory disease in dairy youngstock (www.sciencedirect.com/journal/journal-of-dairy-science).
- The ProYoungStock project – a European research study examining the effects of natural rearing and feeding strategies on health, welfare and economic performance (https://www.proyoungstock.net/about.html).

10 References

Abuelo, A. (2020). Symposium review: late-gestation maternal factors affecting the health and development of dairy calves. *Journal of Dairy Science* 103(4): 3882-3893.

Adler, F., Christley, R. and Campe, A. (2019). Examining farmers' personalities and attitudes as possible risk factors for dairy cattle health, welfare, productivity and farm management: a systematic scoping review. *Journal of Dairy Science* 102: 3805-3824.

Archbold, H., Shalloo, L., Kennedy, E., Pierce, K. M. and Buckley, F. (2012). Influence of age, body weight and body condition score before mating start date on the pubertal rate of maiden Holstein-Friesian heifers and implications for subsequent cow performance and profitability. *Animal* 6(7): 1143-1151.

Bach, A. (2011). Associations between several aspects of heifer development and dairy cow survivability to second lactation. *Journal of Dairy Science* 94(2): 1052-1057.

Bagnella, C. A. and Bartol, F. F. (2019). Relaxin and the 'Milky Way': the lactocrine hypothesis and maternal programming of development. *Molecular and Cellular Endocrinology* 487: 18-23.

Banchero, G. E., Quintas, G., Martin, G. B., Milton, J. T. and Lindsay, D. R. (2004). Nutrition and colostrum production in sheep. 2. Metabolic and hormonal responses to different energy sources in the final stages of pregnancy. *Reproduction, Fertility, and Development* 16(6): 645-653.

Barker, D. J., Winter, P. D., Osmond, C., Margetts, B. and Simmonds, S. J. (1989). Weight in infancy and death from ischaemic heart disease. *Lancet* 2(8663): 577-580.

Barrier, A. C., Dwyer, C. M., Macrae, A. I. and Haskell, M. J. (2012). Short communication: survival, growth to weaning, and subsequent fertility of live-born dairy heifers after a difficult birth. *Journal of Dairy Science* 95(11): 6750-6754.

Barrier, A. C., Haskell, M. J., Birch, S., Bagnall, A., Bell, D. J., Dickinson, J., Macrae, A. I.. and Dwyer, C. M. (2013). The impact of dystocia on dairy calf health, welfare, performance and survival. *The Veterinary Journal* 195(1): 86-90.

Beam, A. L., Lombard, J. E., Kopral, C. A., Garber, L. P., Winter, A. L., Hicks, J. A. and Schlater, J. L. (2009). Prevalence of failure of passive transfer of immunity in newborn heifer calves and associated management practices on US dairy operations. *Journal of Dairy Science* 92(8): 3973-3980.

Beaver, A., Meagher, R. K., von Keyserlingk, M. A. G. and Weary, D. M. (2019). Invited review: a systematic review of the effects of early separation on dairy cow and calf health. *Journal of Dairy Science* 102(7): 5784-5810.

Berry, D. P. and Cromie, A. R. (2009). Associations between age at first calving and subsequent performance in Irish spring-calving Holstein-Friesian dairy cows. *Livestock Science* 123(1): 44-54.

Bland, I., Lang, P. and Hill, J. (2007). The consequences of offering diets high in rumen undegradable protein during the dry period on immunoglobulin G concentrations in serum of neonates, mature dairy cows and colostrum. In: *Proceedings of the 3rd Australasian Dairy Science Symposium*, Melbourne, pp. 378-386.

Boersma, S.-J., Cannas da Silva, J., Mee, J. and Noordhuizen, J. P. T. M. (2010). Diseases of young stock. In: *Farm Health and Productivity Management of Dairy Young Stock*, Boersma, S.-J., Cannas da Silva, J., Mee, J. and Noordhuizen, J. P. T. M. (Eds), Wageningen Academic Publishers, Wageningen, pp. 131-192.

Boland, T. M., Keane, N., Nowakowski, P., Brophy, P. O. and Crosby, T. F. (2005). High mineral and vitamin E intake by pregnant ewes lowers colostral immunoglobulin G absorption by the lamb. *Journal of Animal Science* 83(4): 871-878.

Boulton, A. C., Rushton, J. and Wathes, D. C. (2017). An empirical analysis of cost of rearing dairy heifers from birth to first calving and the time taken to repay these costs. *Animal* 11(8): 1372-1380.

Brickell, J. S., McGowan, M. M. and Wathes, D. C. (2009). Effect of management factors and blood metabolites during the rearing period on growth in dairy heifers on UK farms. *Domestic Animal Endocrinology* 36(2): 67-81.

Britney, J. B., Martin, S. W., Stone, J. B. and Curtis, R. A. (1984). Analysis of early calf-hood health status and subsequent dairy herd survivorship and productivity. *Preventive Veterinary Medicine* 3(1): 45-52.

Bull, R., Ross, R., Olson, D., Blecha, F. and Curtis, S. (1978). Effects of maternal protein restriction on blood composition and pathologic lesions in neonatal calves. *Journal of Animal Science* 47(Suppl. 1): 173.

Burton, J. H., Hosein, A. A., Mcmillan, I., Grieve, D. G. and Wilkie, B. N. (1984). Immunoglobulin absorption in calves as influenced by dietary protein intakes of their dams. *Canadian Journal of Animal Science* 64(5): 185-186.

CAFRE (2020). *Heifer Summary Report*. College of Agriculture, Food & Rural Enterprise, Antrim, Northern Ireland.

Carstens, G. E., Johnson, D. E., Holland, M. D. and Odde, K. G. (1987). Effects of prepartum protein nutrition and birth weight on basal metabolism in bovine neonates. *Journal of Animal Science* 65(3): 745-751.

Civra, A., Altomare, A., Francese, R., Donalisio, M., Aldini, G. and Lembo, D. (2019). Colostrum from cows immunised with a veterinary vaccine against bovine rotavirus displays enhanced in vitro anti-human rotavirus activity. *Journal of Dairy Science* 102(6): 4857-4869.

Cole, J., Hutchinson, J., Bickhart, D. and Null, D. (2013). Optimal age at first calving for US dairy cattle. *Journal of Dairy Science* 91(E-Suppl. 2): 289.

Connelly, M., Berry, D., Sayrs, R., Murphy, J., Lornz, I., O'Doherty, E. and Kennedy, E. (2014). Does iodine supplementation of the pre-partum dairy cow diet affect the serum IgG concentrations, iodine and health status of her calf? *Journal of Dairy Science* 8: 5120-5139.

Cooke, R. F. (2019). Effects on animal health and immune function. *Veterinary Clinics of North America: Food Animal Practice* 35(2): 331-341.

Corah, L. R., Dunn, T. G. and Kaltenbach, C. C. (1975). Influence of prepartum nutrition on the reproductive performance of beef females and the performance of their progeny. *Journal of Animal Science* 41(3): 819-824.

Correa, M. T., Curtis, C. R., Erb, H. N. and White, M. E. (1988). Effect of calf-hood morbidity on age at first calving in New York Holstein herds. *Preventive Veterinary Medicine* 6(4): 253-262.

Croyle, S. L., Belage, E., Khosa, D. K., LeBlanc, S. J., Haley, D. B. and Kelton, D. F. (2019). Dairy farmers' expectations and receptivity regarding animal welfare advice: a focus group study. *Journal of Dairy Science* 102(8): 7385-7397.

Cuttance, E. L., Mason, W. A., Laven, R. A. and Phyn, C. V. C. (2018). The relationship between failure of passive transfer and mortality, farmer-recorded animal health events and body weights of calves from birth until 12 months of age on pasture-based, seasonal calving dairy farms in New Zealand. *The Veterinary Journal* 236: 4-11.

Davis Rincker, L. E., VanderHaar, M. J., Wolf, C. A., Liesman, J. S., Chapin, L. T. and Weber Nielsen, M. S. (2011). Effects of intensified feeding of heifer calves on growth, pubertal age, calving age, milk yield, and economics. *Journal of Dairy Science* 94(7): 3554-3567.

Dawkins, M. (2019). Animal welfare as preventative medicine. *Animal Welfare* 28(2): 137-141.

Dechow, C. (2020). "Rightsize" replacement inventories. *Hoard's Dairyman* 165: 140.

DeNise, S. K., Robison, J. D., Stott, G. H. and Armstrong, D. V. (1989). Effects of passive immunity on subsequent production in dairy heifers. *Journal of Dairy Science* 72(2): 552-554.

De Vries, A. and Marcondes, M. I. (2020). Review: overview of factors affecting productive lifespan of dairy cows. *Animal* 14(S1): 155-164.

Donovan, G. A., Dohoo, I. R., Montgomery, D. M. and Bennett, F. L. (1998). Calf and disease factors affecting growth in female Holstein calves in Florida, USA. *Preventive Veterinary Medicine* 33(1-4): 1-10.

Drackley, J., Pollard, B., Dann, H. and Stamey, J. (2007). First-lactation milk production for cows fed control or intensified milk replacer programs as calves. *Journal of Dairy Science* 1: 614.

Dubrovsky, S., Van Eenennaam, A., Aly, S., Karle, B., Rossitto, P., Overton, M. and Lehenbauer, T. and Fadel, J. (2020). Preweaning cost of bovine respiratory disease (BRD) and cost-benefit of implementation of preventative measures in calves on California dairies: The BRD 10K study. *Journal of Dairy Science* 103: 1583-1597.

Dunn, A., Ashfield, A., Earley, B., Welsh, M., Gordon, A., McGee, M. and Morrison, S. J. (2017). Effect of concentrate supplementation during the dry period on colostrum quality and effect of colostrum feeding regimen on passive transfer of immunity, calf health and performance. *Journal of Dairy Science* 100(1): 357-370.

Eastham, N. T., Coates, A., Cripps, P., Richardson, H., Smith, R. and Oikonomou, G. (2018). Associations between age at first calving and subsequent lactation performance in UK Holstein and Holstein-Friesian dairy cows. *PLoS ONE* 13(6): e0197764.

Ebert, A., Heyn, E., Erhard, M. and Klee, W. (2007). Influence of method of colostrum application on IgG status, incidence of diseases, and weight gain in neonatal calves. In: *Proceedings of the 40th AABP*, p. 221.

Enjalbert, F. (2009). The relationship between trace elements stouts and health of calves. *Revue de Medecine Veterinaire* 160: 429-435.

Ettema, J. F. and Santos, J. E. P. (2004). Impact of age at calving on lactation, reproduction, health, and income in first-parity Holsteins on commercial farms. *Journal of Dairy Science* 87: 2730-2742.

Faber, S. N., Faber, N. E., McCauley, T. C. and Ax, R. L. (2005). Effects of colostrum ingestion on lactation performance. *Professional Animal Scientist* 21(5): 420-425.

Fenlon, C., O'Grady, L., Mee, J. F., Butler, S. T., Doherty, M. L. and Dunnion, J. (2017). A comparison of 4 predictive models of calving assistance and difficulty in dairy heifers and cows. *Journal of Dairy Science* 100(12): 9746-9758.

Franklin, S. T., Newman, M. C., Newman, K. E. and Meek, K. I. (2005). Immune parameters of dry cows fed mannan oligosaccharide and subsequent transfer of immunity to calves. *Journal of Dairy Science* 88(2): 766-775.

Funston, R. N., Larson, D. M. and Vonnahme, K. A. (2010). Effects of maternal nutrition on conceptus growth and offspring performance: implications for beef cattle production. *Journal of Animal Science* 88(13 Suppl.): E205-E215.

Furman-Fratczak, K., Rzasa, A.. and Stefaniak, T. (2011). The influence of colostral immunoglobulin concentration in heifer calves' serum on their health and growth. *Journal of Dairy Science* 94(11): 5536-5543.

Garnsworthy, P. (2012). The contribution of youngstock to production efficiency and environmental impact of dairy systems. In: *Proceedings of the XXVII World Buiatrics Congress*, Lisbon, Portugal, pp. 161-164.

Geiger, A. J., Parsons, C. L. M., James, R. E. and Akers, R. M. (2016). Growth, intake, and health of Holstein heifer calves fed an enhanced preweaning diet with or without postweaning exogenous estrogen. *Journal of Dairy Science* 99(5): 3995-4004.

Gilmore, J. and Mee, J. F. (2012). Transition period management of the suckler cow and calf. *Journal of the Irish Grassland Association* 43: 143-156.

Gleeson, D. and O'Brien, B. (2012). Effect of milk feed source, frequency of feeding and age at turnout on calf performance, live-weigh at mating and 1st lactation milk production. *Irish Veterinary Journal* 65(1): 18.

Gleeson, D. E., O'Brien, B. and Mee, J. F. (2007). Effect of restricting silage feeding prepartum on time of calving, dystocia and stillbirth in Holstein-Friesian cows. *Irish Veterinary Journal* 60(11): 667-671.

Godden, S. M., Haines, D. M., Konkol, K. and Peterson, J. (2009). Improving passive transfer of immunoglobulins in calves. II: Interaction between feeding method and volume of colostrum fed. *Journal of Dairy Science* 92(4): 1758-1764.

Gotowiecka, M., Nicpon, J., Twardon, J., Ochota, M. and Gotowiecka, M. (2012). The influence of alkalosis and acidosis in cows on the content of colostrum and the osmolarity of the gastric juice and blood in calves. In: *Proceedings of the Abstract Book of the XXVII World Buiatrics Congress*, Lisbon, Portugal, p. 208.

Gulliksen, S. M., Lie, K. I., Solverod, L. and Osteras, O. (2008). Risk factors associated with colostrum quality in Norwegian dairy cows. *Journal of Dairy Science* 91(2): 704-712.

Guyot, H., Spring, P., Andrieu, S. and Rollin, F. (2007). Comparative responses to sodium selenite and organic selenium supplements in Belgian Blue cows and calves. *Livestock Science* 111(3): 259-263.

Hall, J. A., Bobe, G., Vorachek, W. R., Estill, C. T., Mosher, W. D., Pirelli, G. J. and Gamroth, M. (2014). Effect of supranutritional maternal and colostral selenium supplementation on passive absorption of immunoglobulin G in selenium-replete dairy calves. *Journal of Dairy Science* 97(7): 4379-4391.

Hamilton, T. and Giesen, L. (1996). Effect of cow condition on calving, cow and calf behaviour, colostrum quality and rebreeding. Agriculture and Agri-Food Canada. *Ontario Beef Research Update*, pp. 30-31.

Hammer, C., Vonnahme, K., Taylor, J., Redmer, D., Luthar, J., Neville, L., Reed, J., Caton, J. and Reynolds, L. (2007). Effects of maternal nutrition and selenium supplementation on absorption of IgG and survival of lambs. *Journal of Dairy Science* 90(Suppl.1): 391.

Hammon, H. M., Liermann, W., Frieten, D. and Koch, C. (2020). Review: importance of colostrum supply and milk feeding intensity on gastrointestinal and systemic development in calves. *Animal* 14(S1): s133-s143.

Haworth, G. M., Tranter, W. P., Chuck, J. N., Cheng, Z. and Wathes, D. C. (2008). Relationships between age at first calving and first lactation milk yield, and lifetime productivity and longevity in dairy cows. *Veterinary Record* 162(20): 643-647.

Hayes, C. J., McAloon, C. G., Carty, C. I., Ryan, E. G., Mee, J. F. and O'Grady, L. (2019). The effect of growth rate on reproductive outcomes in replacement dairy heifers in seasonally calving, pasture-based systems. *Journal of Dairy Science* 102(6): 5599–5611.

Heinrichs, A. J. and Heinrichs, B. S. (2011). A prospective study of calf factors affecting first-lactation and lifetime milk production and age of cows when removed from the herd. *Journal of Dairy Science* 94(1): 336–341.

Heinrichs, A. J., Heinrichs, B. S., Harel, O., Rogers, G. W. and Place, N. T. (2005). A prospective study of calf factors affecting age, body size, and body condition score at first calving of Holstein dairy heifers. *Journal of Dairy Science* 88(8): 2828–2835.

Hennessy, T. and Heanue, K. (2012). Quantifying the effect of discussion group membership on technology adoption and farm profit on dairy farms. *Journal of Agricultural Education and Extension* 18(1): 41–54.

Hogan, I., Doherty, M., Fagan, J., Kennedy, E., Connelly, M., Brady, P., Ryan, C. and Lorenz, I. (2015). Comparison of rapid laboratory tests for failure of passive transfer in the bovine. *Irish Veterinary Journal* 68(1): 18.

Holland, M., Speer, N., Le Fever, D., Taylor, R., Field, T. and Odde, K. (1993). Factors contributing to dystocia due to fetal malpresentation in beef cattle. *Theriogenology* 39: 898–908.

Hook, T. E., Odde, K. G., Aguilar, A. A. and Olson, J. D. (1989). Protein effects on fetal growth, colostrum and calf immunoglobulins and lactation in dairy heifers. *Journal of Animal Science* 67(S1): 539.

Hough, R. L., McCarthy, F. D., Kent, H. D., Eversole, D. E. and Wahlberg, M. L. (1990). Influence of nutritional restriction during late gestation on production measures and passive immunity in beef cattle. *Journal of Animal Science* 68(9): 2622–2627.

Hu, W., Hill, T. M., Dennis, T. S., Suarez-Mena, F. X., Aragona, K. M., Quigley, J. D. and Schlotterbeck, R. L. (2020). Effects of milk replacer feeding rates on growth, performance of Holstein dairy calves to 4 months of age, evaluated via a meta-analytical approach. *Journal of Dairy Science* 103(3): 2217–2232.

Hultgren, J., Svensson, C., Maizon, D. O. and Oltenacu, P. A. (2008). Rearing conditions, morbidity and breeding performance in dairy heifers in southwest Sweden. *Preventive Veterinary Medicine* 87(3–4): 244–260.

Hunter, A. (2015). *Association of Serum Calcium Status at Calving on Survival, Health and Performance of Post-Partum Holstein Cows and Calves*. MSc, Ohio State University.

Hyde, R. M., Green, M. J., Sherwin, V. E., Hudson, C., Gibbons, J., Forshaw, T., Vickers, M. and Down, P. M. (2020). Quantitative analysis of calf mortality in Great Britain. *Journal of Dairy Science* 103(3): 2615–2623.

Illek, J. (2010). The effect of organic and inorganic selenium supplementation on selenium concentrations in colostrum and blood of beef heifers and their calves. In: *Proceedings of the XXVI World Buiatrics Congress*, Chile, p. 89.

Jawor, P., Krol, D., Mee, J. F., Soltysiak, Z., Dzimira, S., Larska, M. and Stefaniak, T. (2017). Infection exposure, detection and causes of death in perinatal mortalities in Polish dairy herds. *Theriogenology* 103: 130–136.

Khan, M. A., Weary, D. M. and von Keyserlingk, M. A. (2011). Effects of milk ration on solid food intake, weaning and performance in dairy heifers. *Journal of Dairy Science* 94(3): 1071–1081.

Larson, D. M., Martin, J. L., Adams, D. C. and Funston, R. N. (2009). Winter grazing system and supplementation during late gestation influence performance of beef cows and steer progeny. *Journal of Animal Science* 87(3): 1147–1155.

Leyan, V., Wittwer, F., Contreras, P. and Kruze, J. (2004). Serum colostrum immunoglobulin concentrations from selenium deficient cows and in the blood of their calves. *Archivos de Medicina Veterinaria* 36: 155-162.

Lin, C. Y., McAllister, A. J., Batra, T. R., Lee, A. J., Roy, G. L., Vesely, J. A., Wauthy, J. M. and Winter, K. A. (1988). Effects of early and late breeding of heifers on multiple lactation performance of dairy cows. *Journal of Dairy Science* 71(10): 2735-2743.

Lombard, J. E., Garry, F. B., Tomlinson, S. M. and Garber, L. P. (2007). Impacts of dystocia on health and survival of dairy calves. *Journal of Dairy Science* 90(4): 1751-1760.

Lora, I., Barberio, A., Contiero, B., Paparella, P., Bonfanti, L., Brscic, M., Stefani, A. L. and Gottardo, F. (2018). Factors associated with passive immunity transfer in dairy calves: combined effect of delivery time, amount and quality of the first colostrum meal. *Animal* 12(5): 1041-1049.

Lorenz, I., Mee, J. F., Earley, B. and More, S. J. (2011). Calf health from birth to weaning. I. General aspects of disease prevention. *Irish Veterinary Journal* 64(1): 10.

Maccari, P., Wiedemann, S., Kunz, H. J., Piechotta, M., Sanftleban, P. and Kaske, M. (2015). Effects of two different rearing protocols for Holstein bull calves in the first 3 weeks of life on health status, metabolism and subsequent performance. *Journal of Animal Physiology and Animal Nutrition* 99(4): 737-746.

Mahmoud, M., Yin, T., Brugemann, K. and Konig, S. (2017). Phenotypic, genetic and single nucleotide polymorphism marker associations between calf diseases and subsequent performance and disease occurrences of first-lactation German Holstein cows. *Journal of Dairy Science* 100(3): 2017-2031.

Mann, S., Leal Yepes, F., Overton, T. and Nydam, D. (2013). The influence of different nutritional planes in the dry period on immunoglobulin G concentration of bovine colostrum. In: *Proceedings of the 46th AABP Conference*, Milwaukee, Wisconsin, p. 170.

McAloon, C., Doherty, M., Donlon, J., Lorenz, I., Meade, J., O'Grady, L. and Whyte, P. (2016). Microbial contamination of colostrum on Irish dairy farms. *Veterinary Record* 1-3. doi: 10/1136/vr.103641.

McCorquodale, C. E., Sewalem, A., Miglior, F., Kelton, D., Robinson, A., Koeck, A. and Leslie, K. E. (2013). Short communication: Analysis of health and survival in a population of Ontario Holstein heifer calves. *Journal of Dairy Science* 96(3): 1880-1885.

McGee, M., Drennan, M. and Caffrey, P. (2006). Effect of age and nutrient restriction pre partum on beef suckler cow serum immunoglobulin concentrations, colostrum yield, composition and immunoglobulin concentration and immune status of their progeny. *Irish Journal of Agricultural and Food Research* 45: 157-171.

Mee, J. F. (2004). The role of micronutrients in bovine periparturient problems. *Cattle Practice* 12: 95-108.

Mee, J. F. (2008). Prevalence and risk factors for dystocia in dairy cattle: a review. *The Veterinary Journal* 176(1): 93-101.

Mee, J. F. (2013). Why do so many calves die on modern dairy farms and what can we do about calf welfare in the future? *Animals* 3(4): 1036-1057.

Mee, J. F. (2014). Impacts of nutrition pre-calving on periparturient dairy cow health and neonatal calf health. *Recent Advances in Animal Nutrition* 45: 37-59.

Mee, J. F. (2018). Intensive care of the newborn dairy calf - knowledge into practice. In: *Proceedings of the 30th World Buiatrics Congress*, Sapporo, Japan, pp. 85-89.

Mee, J. F. (2020). Invited review: denormalising poor dairy youngstock management - dealing with 'farm-blindness'. *Journal of Animal Science* 98(S1): 140-S149.

Mee, J. F. and Kenneally, J. (2017). Why Do Calves Die at Calving and What Can You Do about It? In: *Irish Dairying Resilient Technologies*, Butler, S., Horan, B., Mee, J. F. and Dillon, P. (Eds), Teagasc, Moorepark, Ireland, pp. 150–151.

Mee, J. F. and Boyle, L. (2020). Invited review: Assessing whether dairy cow welfare is 'better' in pasture- than in confinement-based management systems. *New Zealand Veterinary Journal* 68(3): 168–177.

Mee, J. F., Berry, D. P. and Cromie, A. R. (2011). Risk factors for calving assistance and dystocia in pasture-based Holstein-Friesian heifers and cows in Ireland. *The Veterinary Journal* 187(2): 189–194.

Mee, J. F., Geraghty, T., O'Neill, R. and More, S. J. (2012). Bioexclusion of diseases from dairy and beef farms: risks of introducing infectious agents and risk reduction strategies. *The Veterinary Journal* 194(2): 143–150.

Mee, J. F., Grant, J., Sanchez-Miguel, C. and Doherty, M. (2013b). Pre-calving and calving management practices in dairy herds with a history of high or low bovine perinatal mortality. *Animals* 3(3): 866–881.

Mee, J. F., McCarthy, M. and O'Grady, L. (2018). Contract rearing - a disaster waiting to happen or a fundamental component of herd expansion? *Irish Veterinary Journal* 8: 474–476.

Mee, J. F., O'Farrell, K. J., Reitsma, P. and Mehra, R. (1996). Effect of a whey protein concentrate used as a colostrum substitute or supplement on calf immunity, weight gain and health. *Journal of Dairy Science* 79(5): 886–894.

Mee, J. F., Sanchez-Miguel, C. and Doherty, M. (2013a). An international Delphi study of the causes of death and the criteria used to assign cause of death in bovine perinatal mortality. *Reproduction in Domestic Animals* 48(4): 651–659.

Mee, J. F., Sanchez-Miguel, C. and Doherty, M. (2014). Influence of modifiable risk factors on the incidence of stillbirth/perinatal mortality in dairy cattle. *The Veterinary Journal* 199(1): 19–23.

Miller, M. and Faust, M. (2000). Effects of performance and physiological characteristics of dairy heifers on first lactation yield and lifetime performance. *Journal of Dairy Science* 83(Suppl. 1): 230.

Moallem, U., Werner, D., Lehrer, H., Zachut, M., Livshitz, L., Yakoby, S. and Shamay, A. (2010). Long-term effects of ad libitum whole milk feeding prior to weaning and prepubertal protein supplementation on skeletal growth rate and first-lactation milk production. *Journal of Dairy Science* 93(6): 2639–2650.

Mock, T., Mee, J. F., Dettwiler, M., Rodriguez-Campos, S., Husler, J., Michel, B., Hafliger, I. M., Drogemuller, C., Bodmer, M. and Hirsbrunner, G. (2020). Evaluation of an investigative model in dairy herds with high calf perinatal mortality rates in Switzerland. *Theriogenology* 148: 48–59.

Modh Nor, N., Steenveld, W., Mourits, M. C. and Hogeveen, H. (2012). Estimating the costs of rearing young dairy cattle in the Netherlands using a simulation model that accounts for uncertainty related to diseases. *Preventive Veterinary Medicine* 106(3-4): 214–224.

Mohd Nor, N., Steeneveld, W., van Werven, T., Mourits, M. C. and Hogeveen, H. (2013). First-calving age and first lactation milk production on Dutch dairy farms. *Journal of Dairy Science* 96(2): 981–992.

More, S. J., Doherty, M. L. and O'Grady, L. (2017). An investigative framework to facilitate epidemiological thinking during herd problem-solving. *Irish Veterinary Journal* 70: 11.

Morgans, L., Bolt, S., van Dijk, L., Reyher, K. and Main, D. (2018). Farmer action groups – A participatory approach to instigating change on farm. In: *Proceedings of the 30th WBC*, Sapporo, Japan, p. 132.

Moriarty, J., Toolan, D., Casey, M., Barry, J. and McLaughlin, J. (2007). Stunting in cattle on three farms associated with calf-hood diseases. *Irish Veterinary Journal* 60: 742-746.

Morin, D. E., Nelson, S. V., Reid, E. D., Nagy, D. W., Dahl, G. E. and Constable, P. D. (2010). Effect of calving and management factors on colostral IgG concentration in dairy cows. *Journal of the American Veterinary Medical Association* 237(4): 420-428.

Morrill, K. M., Marston, S. P., Whitehouse, N. L., Van Amburgh, M. E., Schwab, C. G., Haines, D. M. and Erickson, P. S. (2010). Anionic salts in the prepartum diet and addition of sodium bicarbonate to colostrum replacer, and their effects on immunoglobulin G absorption in the neonate. *Journal of Dairy Science* 93(5): 2067-2075.

Morrison, S., Carson, A., Matthews, D., Kilpatrick, D., McCluggage, I. and Mulholland, M. (2010). Latest research on calf management for performance. In: *Proceedings of the AgriSearch Seminar on Dairy Genetics and Heifer Management*, Hillsborough, Northern Ireland, pp. 68-98.

Morrison, S., Scoley, G. and Barley, J. (2013). The impact of calf health on future performance. *Irish Veterinary Journal* 3: 264-268.

Morrison, S. J., Wicks, H. C., Carson, A. F., Fallon, R. J., Twigge, J., Kilpatrick, D. J. and Watson, S. (2012). The effects of calf nutrition on the performance of dairy herd replacements. *Animal* 6(6): 909-919.

Moscuzza, C., Milicich, H., Alvarez, G., Gutierrez, B. and Nahum, M. (2014). Calving assistance influences the occurrence of umbilical cord pathologies treated surgically in calves. *Turkish Journal of Veterinary and Animal Sciences* 38: 405-408.

Mousa, M., Seykora, A., Abdellatif, M. and Mousa, E. (2015). Genetic and phenotypic correlations of age at first calving with some calf-hood fitness and health traits in Holstein heifers. *Egyptian J. Animal Production* 52(Suppl. Issue): 37-43.

Nakao, T., Zhang, W., Kida, K., Moriyoshi, M. and Nakada, K. (2000). Effects of nutrition during gestation on birth weight and viability of calves, and weight and expulsion time of placenta in dairy cattle. In: *Proceedings of the 33rd Annual Conference of the American Association of Bovine Practitioners, Rapid City, South Dakota, United States of America*.

Nejedla, E., Stanek, S., Zouharova, M., Fleischer, P., Nedbalcova, K. and Slosarkova, S. (2019). Microbial quality of colostrum from Czech dairy herds. *Cattle Practice* May, p. 48.

Nowak, W., Mikula, R., Kasprowicz-Potocka, M., Ignatowicz, M., Zachwieja, A., Paczynska, K. and Pecka, E. (2012a). Effect of cow nutrition in the far-off period on colostrum quality and immune response of calves. *Bulletin of the Veterinary Institute in Pulawy* 56(2): 241-246.

Nowak, W., Mikula, R., Zachwieja, A., Paczynska, K., Pecka, E., Drzazga, K. and Slosarz, P. (2012b). The impact of cow nutrition in the dry period on colostrum quality and immune status of calves. *Polish Journal of Veterinary Sciences* 15(1): 77-82.

Overton, M. W. and Dhuyvetter, K. C. (2020). An abundance of replacement heifers: what is the economic impact of raising more than are needed? *Journal of Dairy Science* 103(4): 3828-3837.

Pereira, R., Siler, J., Sischo, W., Davis, M. and More, D. (2014). Intra-farm communication about calf health on US dairy farms. In: *Proceedings of the 28th WBC*, Cairns, Australia, p. 185.

Perry, V. E. A., Copping, K. J., Miguel-Pacheco, G. and Hernandez-Medrano, J. (2019). The effects of developmental programming upon neonatal mortality. *Veterinary Clinics of North America: Food Animal Practice* 35(2): 289–302.

Quigley, J. D. and Drewry, J. J. (1998). Nutrient and immunity transfer from cow to calf pre- and postcalving. *Journal of Dairy Science* 81(10): 2779–2790.

Quigley, J. D., Lago, A., Chapman, C., Erickson, P. and Polo, J. (2013). Evaluation of the Brix refractometer to estimate immunoglobulin G concentration in bovine colostrum. *Journal of Dairy Science* 96(2): 1148–1155.

Raeth-Knight, M., Chester-Jones, H., Hayes, S., Linn, J., Larson, R., Ziegler, D., Ziegler, B. and Broadwater, N. (2009). Impact of conventional or intensive milk replacer programs on Holstein heifer performance through six months of age and during first lactation. *Journal of Dairy Science* 92(2): 799–809.

Renaud, D. L., Waalderbos, K. M., Beavers, L., Duffield, T. F., Leslie, K. E. and Windeyer, M. C. (2020). Risk factors associated with failed transfer of passive immunity in male and female dairy calves: a 2008 retrospective cross-sectional study. *Journal of Dairy Science* 103(4): 3521–3528.

Robison, J. D., Stott, G. H. and DeNise, S. K. (1988). Effects of passive immunity on growth and survival in the dairy heifer. *Journal of Dairy Science* 71(5): 1283–1287.

Rytkonen, A., Hanninen, L. and Manninen, M. (2004). Effect of restricted feeding on the liveweight of suckler cows and their calves, quality of colostrum and calf serum IgG concentration. *Suomen Elainlaakarilehti* 110: 427–433.

Santos, J. E., DePeters, E. J., Jardon, P. W. and Hubert, J. T. (2001). Effect of prepartum dietary protein level on performance of primigravid and multiparous Holstein dairy cows. *Journal of Dairy Science* 84(1): 213–224.

Shalloo, L., Cromie, A. and McHugh, N. (2014). Effect of fertility on the economics of pasture-based dairy systems. *Animal* 8(Suppl. 1): 222–231.

Shamay, A., Werner, D., Moallem, U., Barash, H. and Bruckental, I. (2005). Effect of nursing management and skeletal size at weaning on puberty, skeletal growth rate and milk production during first lactation of dairy heifers. *Journal of Dairy Science* 88(4): 1460–1469.

Shell, T. M., Early, R. J., Carpenter, J. R. and Buckley, B. A. (1995). Prepartum nutrition and solar radiation in beef cattle: 2. Residual effects on postpartum milk yield, immunoglobulin, and calf growth. *Journal of Animal Science* 73(5): 1303–1309.

Shivley, C. B., Lombard, J. E., Urie, N. J., Haines, D. M., Sargent, R., Kopral, C. A., Earlywine, T. J., Olson, J. D. and Garry, F. B. (2018). Preweaned heifer management on US dairy operations: Part II. Factors associated with colostrum quality and passive transfer status of dairy heifer calves. *Journal of Dairy Science* 101(10): 9185–9198.

Silper, B., Madureira, A., Burnett, T., de Passeille, A., Rushen, J. and Cerri, R. (2013). Relationships between pre- and postweaning growth on estrus behavior and reproductive parameters of Holstein heifers. *Journal of Dairy Science* 91(E-Suppl. 2): 671.

Soberon, F., Raffrenato, E., Everett, R. W. and Van Amburgh, M. E. (2012). Preweaning milk replacer intake and effects on long-term productivity of dairy calves. *Journal of Dairy Science* 95(2): 783–793.

Soberon, F. and Van Amburgh, M. E. (2013). The effect of nutrient intake from milk or milk replacer of preweaned dairy calves on lactation milk yield as adults: a meta-analysis of current data. *Journal of Animal Science* 91(2): 706–712.

Stanton, A. L., Kelton, D. F., LeBlanc, S. J., Wormuth, J. and Leslie, K. E. (2012). The effect of respiratory disease and a preventative antibiotic treatment on growth, survival, age at first calving, and milk production of dairy heifers. *Journal of Dairy Science* 95(9): 4950-4960.

Steele, M. (2020). Age at first calving in dairy cows: which months do you aim for to maximise productivity? *Veterinary Evidence* 5(1): 1-22.

Stockdale, C. and Smith, C. (2004). Effect of energy and protein nutrition in late gestation on immunoglobulin G in the colostrum of dairy cows with varying body condition scores. *Animal Production in Australia* 25: 176-179.

Svensson, C. and Hultgren, J. (2006). The effect of calf rearing factors on first-lactation milk production. In: *Proceedings of the World Buiatrics Congress*, Nice, France, CD-ROM, OS37-1.

Svensson, C. and Hultgren, J. (2008). Associations between housing, management, and morbidity during rearing and subsequent first-lactation milk production of dairy cows in Southwest Sweden. *Journal of Dairy Science* 91(4): 1510-1518.

Taylor, P. D. and Poston, L. (2007). Developmental programming of obesity in mammals. *Experimental Physiology* 92(2): 287-298.

Terre, M., Tejero, C. and Bach, A. (2009). Long-term effects on heifer performance of an enhanced-growth feeding programme applied during the preweaning period. *Journal of Dairy Research* 76(3): 331-339.

Vaarst, M. and Sorensen, J. T. (2009). Danish farmers' perceptions and attitudes related to calf management in situations of high versus no calf mortality. *Preventive Veterinary Medicine* 89(1-2): 128-133.

Van der Fels-Klerx, H. J., Martin, S. W., Nielen, M. and Huirne, R. B. M. (2002). Effects on productivity and risk factors of bovine respiratory disease in dairy heifers: a review for the Netherlands. *NJAS – Wageningen Journal of Life Sciences* 50(1): 27-45.

Van Hese, I., Goossens, K., Vandaele, L. and Opsomer, G. (2020). Invited review: microRNAs in bovine colostrum – focus on their origin and potential health benefits for the calf. *Journal of Dairy Science* 103(1): 1-15.

Waltner-Toews, D., Martin, S. W. and Meek, A. H. (1986). The effects of early calf-hood health status on survivorship and age at first calving. *Canadian Journal of Veterinary Research* 50(3): 314-317.

Warnick, L., Erb, H. and White, M. (1994). The association of calf-hood morbidity with first lactation calving age and dystocia in New York Holstein herds. *Kenya Veterinarian* 18: 177-179.

Warnick, L. D., Erb, H. N. and White, M. E. (1995). Lack of association between calf morbidity and subsequent first lactation milk production in 25 New York Holstein herds. *Journal of Dairy Science* 78(12): 2819-2830.

Warnick, L., Erb, H. and White, M. (1997). The relationship of calf-hood morbidity with survival after calving in 25 New York Holstein herds. *Preventive Veterinary Medicine* 31: 263-266, 268-273.

Williams, D. R., Pithau, P., Garcia, A., Champagne, J., Haines, D. M. and Aly, S. S. (2014). Effect of three colostrum diets on passive transfer of immunity and preweaning health in calves on a California dairy following colostrum management training. *Veterinary Medicine International* 2014: 1-9. Article ID 98741.

Zanton, G. I. and Heinrichs, A. J. (2005). Meta-analysis to assess effect of prepubertal average daily gain of Holstein heifers on first-lactation production. *Journal of Dairy Science* 88(11): 3860-3867.

www.ingramcontent.com/pod-product-compliance
Lightning Source LLC
Chambersburg PA
CBHW050537270326
41926CB00015B/3266